CLUES
W9-BTD-271

The Aging Brain

MAPS OF THE MIND

Steven Rose, General Editor

MAPS OF THE MIND
Steven Rose, General Editor

Pain: The Science of Suffering
Patrick Wall

The Making of Intelligence
Ken Richardson

How Brains Make Up Their Minds
Walter J. Freeman

Sexing the Brain
Lesley Rogers

Intoxicating Minds: How Drugs Work
Ciaran Regan

How to Build a Mind: Toward Machines with Imagination
Igor Aleksander

The Unbalanced Mind
Julian Leff

The Aging Brain

Lawrence Whalley

Columbia University Press
New York

Columbia University Press
Publishers Since 1893
New York Chichester, West Sussex

Library of Congress Cataloging-in-Publication Data
Whalley, Lawrence J.
The aging brain / Lawrence Whalley.
p. cm. — (Maps of the mind)
Includes bibliographical references and index.
ISBN 0-231-12024-9 (cloth : alk. paper)
1. Brain—Aging. I. Title. II. Series.
QP430 .W48 2001
612.8'2—dc21
2001032380

∞
Casebound editions of Columbia University Press books are printed on permanent and durable acid-free paper.

Printed in Great Britain

c 10 9 8 7 6 5 4 3 2 1

First published by Weidenfeld & Nicolson Ltd., London

For my family

Contents

'The empires of the future will be empires of the mind'
Winston Churchill (1943)

Acknowledgements

I am indebted to my colleagues, particularly Ian Deary and John Starr with whom I have researched and studied for almost fifteen years. I have special thanks for Peter Tallack at Weidenfeld & Nicolson who guided me through the first drafts of this book, producing, at each stage, very substantial improvements of style and clarity. My brother-in-law, Joe Paley, deserves more than a passing mention. His insistence on careful planning and on getting to the point was of much benefit. I also thank Patricia Whalley, who painstakingly read the clumsy first drafts without complaint, and provided much encouragement to complete the task ahead.

I must also thank Steven Rose for inviting me to contribute to the Maps of the Mind series and for his gentle, but persuasive approach to my making much needed changes. If I have achieved clarity in some of the more dense scientific sections, this must be attributable to his patient advice and my active emulation of his lucid style. Nicola Jeanes was very supportive throughout the production process and I will forever be indebted to the meticulous attention paid by an anonymous copy-editor who rendered some of my more convoluted and ill-considered sentences into very plain English. There are others too numerous to list by name, who have all encouraged my own scientific development. Most influential were my teachers in Edinburgh, Ian Oswald, Ralph McGuire and Norman Kreitman. Like many earnest pupils their imaginary presence beside my writing desk encouraged me to consider very carefully whatever claims I was making at the time. Lessons learnt so long ago are perhaps not remembered as accurately as they should be. Whatever mistakes I have made remain my own.

Preface

This short book examines what happens to the brain in the final decades of life. There is nothing new in the simple description of the effects of age on the brain: more detailed accounts can be found in the many textbooks read by health workers who investigate, treat and support the medical problems of old age. What is different is the way in which this book draws on the startling advances in molecular biological research and computer technology to counter the widespread pessimism about what the future has in store for us. These new ideas hold out the possibility of transforming life in old age by preventing or even reversing the damage to brain cells implicated in mental decline.

Recent years have seen an amazing improvement in the general physical health of old people. Nowadays more people enjoy a healthy old age than ever before. If modern medicine can achieve this progress in physical health in just a few years, can similar advances be made in mental health in old age?

Many of the characteristics acquired by old people are widely supposed to result from a single ageing process. There is really very little evidence to support this idea. What used to be regarded as a single process is now known to advance along at least two distinct pathways. One is bodily wear and tear – the trials and tribulations of life. This is the routine stuff of specialists in geriatric medicine, who know that limiting this type of environmental damage can help promote successful ageing. The other is the group of innate processes loosely termed 'biological ageing'. As yet these processes are not amenable to intervention to the same degree; composed largely of inbuilt cellular senescence and damage repair pathways, they are under genetic control.

The exciting prospect held out by the scientific study of the ageing brain is that the brain may be able to compensate for its own ageing. Just as bodily ageing can be divided into 'environmental' and 'genetic ' pathways, so can compensation for brain ageing. Applying the sophisticated techniques of molecular biology to the problem of the ageing brain has explained some rare types of dementia and yielded clues to the catastrophic failure of compensatory repair mechanisms in the most common form of dementia – that arising after the age of seventy-five.

There is a popular belief that much of brain ageing can be attributed to a generalised loss of brain cells with age. Although once widely taught, this dictum – and another, that brain cells lose the ability to replicate – is now seen to be incorrect: loss of brain cells with ageing is not as extensive as once thought, and is certainly not generalised throughout the brain. If brains do not lose cells in the numbers once supposed, and if some surviving cells retain a capacity to replicate, there may be real prospects of slowing or even preventing some of the worst effects of brain ageing.

Another widespread idea about ageing is that mental and physical activity – found preferably in an active social life – is the best way to delay mental decline. Although decline certainly occurs in most old people, there are many who retain – or even improve – their mental abilities across the life span. What can these old people teach us about the rules for successful brain ageing? This topic is perhaps the most immediately rewarding of current studies on the ageing brain. Across the world, research groups have charted the physical and psychological health of middle-aged people well into their ninth decade of life. Lessons so far point to some simple lifestyle rules which, if followed, certainly improve the chances of keeping mental abilities intact: keep fit, keep mentally active, eat a balanced diet, socialise and don't smoke.

Over the past forty years colossal strides have been made in two areas that profoundly affect our future. The first is computer technology, which has changed our lives, made us more productive and is likely to enhance our existence for generations to come. The nature of this existence is the second area of progress: in the space of just two or three generations, industrial societies have established the study of science as the engine of progress towards better health for all.

The biological sciences have become the dynamo pushing the boundaries of health enhancement beyond the gains made by the social improvers of the late nineteenth century. In the twentieth century successive eradications of once lethal diseases marked milestones in the advance of science in the campaign against human disease. Huge progress was made in treating and preventing infectious, vascular, metabolic and inflammatory diseases. Improved surgical techniques and materials conquered previously intractable problems like organ replacement. Even psychiatry had its successes. New drugs reduced the mortality and need for long-term institutional care of many with severe mental illness. The overall effect was to increase life expectancy more rapidly than at any time in human history. More people than ever before are now living into old age.

There is a popular misconception that success in the fight against disease has simply replaced premature infant death or adult disease with untreatable disorders of late life. The truth is quite different. Improved health in pregnancy, childhood and adult life leads to a healthy old age. A greater proportion of old people now enjoy better physical health than ever before. The aim of this book is to help achieve for the brain what the final decades of the last century achieved for the body: fitter old brains for fitter old people.

Chapter 1

What is ageing?

Ageing and the healthy brain

Why do some old people remain alert and vigorous at an age when others are declining mentally and physically? Does their apparent advantage have a biological basis, and if so, could this success be transferred to others predisposed to age more quickly? If this is achievable, does brain ageing then become the last obstacle to extension of our useful life span? This books looks at the answers being generated by current research into brain ageing, and the grounds for optimism.

Brain scientists used to be pessimistic about the prospects of slowing or preventing brain ageing. They saw the brain as a vastly complex organ, with billions of brain cells, kilometres of intricate wiring, and little capacity to repair itself. Such views are now yielding to a different understanding of what happens as the brain ages; what once seemed a simple tale of degeneracy and brain cell death is being replaced by a lifelong story of good 'housekeeping', where brain cells rely on the body's general health and nutrition to maintain complex functions and compensate for the wear and tear of everyday life. Like criminal behaviour, the causes of brain ageing come in many forms; and just as criminals can flourish in particular sections of society, so can the causes of brain ageing be more potent in the brains of susceptible individuals.

Brain ageing is not a simple phenomenon, but a composite of several processes. Biological ageing may be distinguished from the effects of age-related diseases – the cumulative effects of diseases that are more common in old people (that is, they are age-related) – but ageing processes are neither sufficient nor necessary for these diseases to occur (that is, they are not age-dependent).

In this chapter I shall emphasise the 'free radical' hypothesis of ageing, because of all the various ageing theories this seems the most relevant to the brain. The conclusion must be that a single process cannot account for what we see in the ageing brain. We shall also look at the structure of the normal brain and its development.

Ageing and disability

Developed societies are anticipating some alarming social problems caused by the remarkable increase in average life expectancy since 1900 – largely due to better living conditions, better control of childhood infections, and reduced risk of some adult disorders. With each successive generation, a greater proportion survived into old age as the twentieth century progressed, leading professional and political commentators to express concerns about the 'burden' of old people, partly from an economic standpoint but mostly in terms of the physical disabilities of old age. Impaired sight, hearing, mobility and mental function each jeopardise independence. In the United Kingdom in the 1970s, about one person in seven aged eighty years or more was disabled; around twenty times more than those aged about fifty. If the causes of illness in old age remain much the same and these illnesses start at similar ages, then from the age of about sixty we should all expect to spend about 25 per cent of our remaining years with at least one disability.

If this had continued to be true, as more people lived longer, so more would spend more of their old age with disabilities of some kind. Simple predictions like this are now seen as short-sighted and do not reflect what is happening today to the health of old people. Not only has life expectancy improved, but also old people are less disabled. For example, the proportion of old men who had a moderate degree of disability halved in the twenty years after 1975 (Figure 1). The best explanation for what is one of the great unsung successes of modern geriatric and public health medicine is that this period of disability has been compressed into the final year or two of life. The fact that this effect seems most marked among the better educated and the

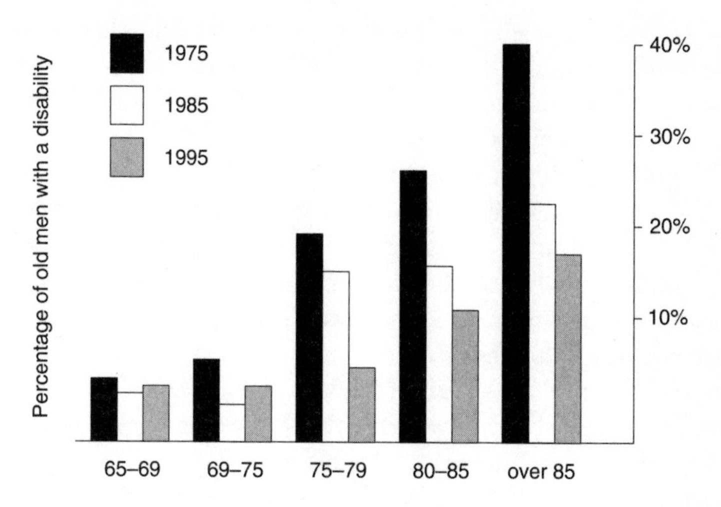

Figure 1 **Old people are becoming healthier.** The proportion of old men unable to perform activities to support their own independence has halved since 1975.

more affluent shows that improvements are not spread evenly among old people. The same factors that predispose the less well-off to greater degrees of cardiovascular disease, respiratory disease and cancer may also make them vulnerable to the chronic disabilities of old age. However, the overall effect has been to postpone premature death, reduce disability, and increase the likelihood that people remain in good health until close to death. Success on this scale should be achievable for other diseases, especially those – like dementia – that affect the brain.

Much of the disease linked to ageing and causing disability in old age may be preventable – or at least its onset may be postponed. This is suggested by differences in age-related disease rates between social classes both within and between countries. In Japan, for example, the link between age and the onset of chronic diseases is weaker than in the UK, and some Mediterranean countries have rates that are intermediate between Japan and the UK. In the USA, cardiovascular deaths have halved since 1970, while they have nearly doubled in Hungary. Over the same period, hip fracture in men has doubled in the UK. These variations establish two facts firmly in the minds of many working to improve the health of old people: first, that the common disabling conditions often thought of as an inevitable consequence of ageing are no such thing at all; and second, that many disabilities related to age are affected by the environment and by public health policy (as in the case of heart disease deaths in the USA). They can be successfully prevented by intervention.

Improvements in the health of old people in developed countries can be attributed to the same causes as improved life expectancy. Even more improvements may result from public health measures to improve diet, reduce the risk of vascular disease and minimise exposure to cancer-causing agents. Currently, it is the better educated and more affluent who adopt healthier lifestyles in greatest numbers – partly because their affluence gives them greater choice about how they live, but perhaps also because they anticipate a more rewarding old age.

Many of the diseases once considered to be simply part of the ageing process – sometimes called the 'degenerative' diseases – are being looked at in a new light as research reveals the effects of the interaction between genetic makeup and the environment, with the consequent potential for further health gains by the growing numbers of old people in our society. One influential opinion is that our life span is largely determined by the genes we inherit from our parents, and variations from this life expectancy are mostly determined by our experiences and lifestyle choices. From this standpoint, what we typically regard as ageing is a composite of at least two distinct sets of factors: one is the consequence of diseases associated with ageing; the other is the outcome of biological processes that impair body functions with increasing age.

The brain does not age in isolation from the body. It is affected by any deterioration in bodily health – especially anything that diminishes its blood supply. Common disorders such as cancer, stroke and diabetes become more frequent in old age. When these diseases arise in middle age they produce low levels of the same type of tissue damage found in otherwise healthy old people. Not surprisingly, this has tempted some scientists to propose that in diseased individuals 'pathological' or 'accelerated' ageing contributes to their chances of developing disease. Studies of very old people do not support this idea. Although some disorders are more common in late life, their incidence declines at the extreme of age. In those unusually old individuals there is often little evidence of chronic disease, but much to suggest that an ageing process is present that is quite independent of disease. The limitation of life span in the absence of disease appears to involve processes that are distinct from disease and probably have their own genetic determinants. These processes are called ageing.

What is ageing?

The study of ageing is full of theories but lacks a bedrock of fact. Ever since ancient alchemists searched for the 'elixir of life' almost all scientists working on the problem of ageing have examined or even proposed a 'unified theory of ageing' of some sort. Once a catalogue of the types of harmful changes found in ageing cells has been drawn up, it seem impossible to resist the idea that these age-related alterations share common features and perhaps even a common cause. The assumption is sometimes made that age-related biological changes are dependent on a single ageing process: but this may not be correct.

The assumption that there is a fundamental biological process labelled 'ageing' responsible for age differences in disease incidence was challenged in 1977 by Oxford professors Richard Doll and Richard Peto. They argued persuasively that it is unnecessary – and potentially misleading – to assume that a single biological process called 'ageing' awaits discovery and holds the key to disorders like cancer and stroke. Their case is based on consideration of the evolutionary pressures that would favour individuals with genes for late-onset disease over those with early-onset genes. This argument leaves room for other biological processes to occur with ageing, but does not assign them a major role in causing age-related disease. An important aspect of the argument developed by Peto and Doll is that there is no good reason to believe the molecular mechanisms leading to age-related disease are necessarily similar to the molecular mechanisms determining the rate of biological ageing – something that will become relevant later when we look at the molecular determinants of dementia.

Biological ageing

Early scientific studies of the biology of ageing gained impetus from cancer research. Likely culprits in both cancer and ageing included the effects of accumulated damage, repeated errors in making key proteins, and fresh genetic mutations. When coupled with defective or overwhelmed repair mechanisms, these gave plenty of scope to explain individual differences in ageing and also the link between cancer and ageing. These theories acquired names such as the 'accumulated damage theory of ageing' or the 'mutagenic theory of ageing'. Each theory has stimulated good-quality research. Some suggest that whatever the nature of the biological process responsible for ageing, it will prove to have genetic components. Potentially, these could influence the life cycle of the cell and compensatory responses to ageing processes. When environmental influences are added to this complex array of factors, the best biological model of ageing becomes indistinguishable from most contemporary causal models of chronic age-related diseases. Not surprisingly, this makes the argument of Peto and Doll all the more attractive for many health researchers in the field of ageing.

At first glance, defining biological ageing might seem to be a simple matter: something grows, reaches an apex, and then begins a period of decline leading to death, which could be called 'ageing'. This type of definition can be applied to the brain. For example, if we were to count the number of brain cells at various stages during our life span, then the brain would reach its apex somewhere around the third year of life and decline thereafter. However, it is clearly nonsense to say that ageing begins so early. We might say instead that in the absence of disease, any age-related decline in the capacity of an organ or body system to adapt to environmental change constitutes ageing. This definition seems to work well; but many biologists who study ageing processes prefer to use terms that indicate the precise biological process under consideration. Some scientists will focus their research on a specific cellular mechanism such as repair of DNA in the cell nucleus. Others concentrate on cell division ('replicative senescence' or 'somatic mutation' theories), or on the functions of an entire body system (such as the immune system), or on the links between nutrition, reproductive behaviour and longevity (behavioural biology). For many scientists, ageing refers to the passage of time and no more than that. The convention among some scientists who study ageing is that the development of sexual maturity conveniently marks the point of onset of biological ageing processes. This is generally true for most organs; but it may not be true for the brain.

The molecular biological revolution

The last ten years of the twentieth century saw dramatic changes in the scientific study of ageing, as the tools of molecular biology became sufficiently exact to prise apart diverse genetic influences on ageing processes. This revolution in a test-tube began in the 1980s with the invention of the polymerase chain reaction (PCR), which made it possible to generate numerous copies of single genes, allowing complex analyses of the functions of each gene and its role in health and disease.

Sometimes the risk of disease is found to be increased when a small section of the genetic code material, DNA, is altered. This is known as a mutation, and may be a simple substitution, a complex insertion, or a subtracton of genetic material. These alterations in DNA occur naturally but can also be produced artificially in the laboratory. This gives the technique its enormous power to unravel the genetic contribution to disease and help us understand the genetic control of body functions in health.

Separation of the components of biological ageing starts at the molecular level. One approach is to look for single gene mutations that alter the life span. So far this 'mutational' approach has proved helpful in just one species – the roundworm, *Caenorhabditis elegans* – though there is some hope that similar genes will be found in other animals.

Theories of biological ageing

At the turn of the nineteenth century, biologists proposed the 'rate of living' hypothesis to explain differences in maximum life span between species. They had observed that animals with the highest metabolic rates had the shortest life spans. Interest in the biological effects of radiation in the 1940s and 1950s led to speculation that radiation damage was caused by the formation in body water of the free radicals hydrogen peroxide and the superoxide anion. (Free radicals are small groups of atoms that carry an unpaired electron, making them extremely reactive

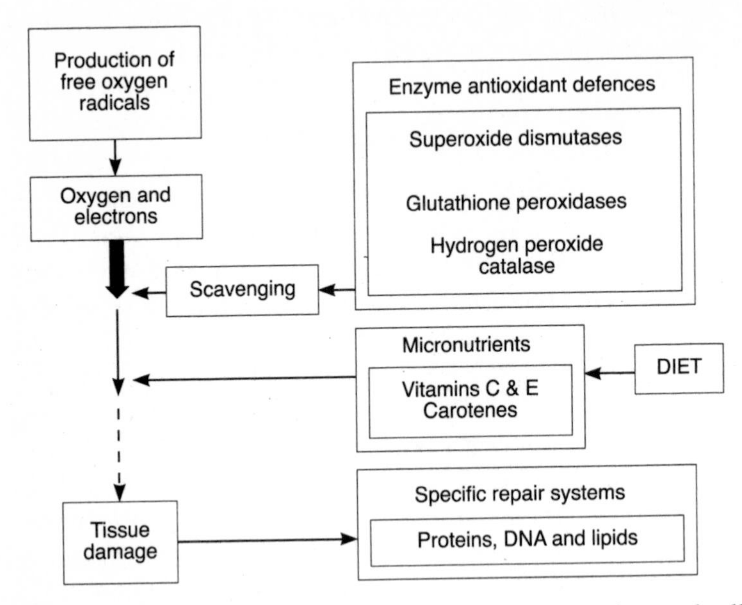

Figure 2 **Free oxygen radicals** are formed in the course of normal cell respiration. They are responsible for a progressive accumulation of cell damage that is linked to age-related diseases. There are two types of antioxidant defence. Intrinsic defences are mostly enzymes and extrinsic defences are compounds taken in diet – mostly fresh fruit and vegetables. Repair systems in place are specifically designed for the type of damage produced.

and thus potentially destructive to body tissues.) In 1956, Denham Harman suggested that free radicals generated during aerobic respiration cause the cumulative damage of ageing. The existence of free radicals was largely hypothetical in the 1950s, but within fifteen years metabolic pathways had been discovered that could generate free radicals. Soon, the body's defences against free radical damage were mapped out. It is now certain that free radicals exist naturally and damage tissue – including proteins, fats and genetic material. Defences against such oxidative damage include an impressive array of antioxidant enzymes as well as some small antioxidant molecules derived from fruit and vegetables in the diet. The relationship between antioxidant generation, antioxidant defence and repair mechanisms is shown in Figure 2.

The free radical theory of ageing is not incompatible with other ageing theories. Tom Kirkwood has put forward the 'disposable soma' theory: this argues that it is potentially disadvantageous to a species to invest too heavily in antioxidant defences when there are more pressing needs to be met to guarantee survival of offspring. The 'somatic mutation' theory holds that ageing is the result of accumulated DNA mutations; this theory is supported by evidence that the life span of a species is linked to its genetic programmes to repair DNA damage. It is not known what proportion of DNA mutations is caused by free radical damage, and the size of this effect is the subject of current intense research.

Another theory of ageing suggests that cells become less efficient at obtaining energy from the metabolism of sugar (glucose). This process (cellular respiration) takes place in specialised structures in the cell, known as mitochondria. These structures contain some DNA derived from the mother and are an important site of free radical production. The dependency of the cell on the mitochondria for energy and to shut away potentially harmful calcium has been seen as one of the body's weak links. The 'mitochondrial' theory of ageing proposes that mutations of mitochondrial DNA contribute to ageing by jeopardising the vitality of the cell. In addition, sites of free radical production are thought to be especially susceptible to free radical damage. This idea neatly ties together the mitochondrial and the free radical theories of ageing, and is attractive because it accounts for a gradual process. Mitochondrial DNA is especially vulnerable because of its association with free radical production and its lack of local protection and repair enzymes. Damage by free radicals could further disrupt cellular respiration, leading to greater free radical production and more DNA damage in a downward spiral of decline.

A widely held view is that a gradual reduction in functional efficiency of cells is at the heart of biological ageing. This reduction is confidently attributed to the unchecked damage to cells by free radicals. The nervous system is often used to study the molecular biology of ageing because brain cells (neurons) do not divide like other body cells and must make good as best they can any damage suffered. Unrepaired damage to neurons causes a recurring cycle of faults in the making of proteins, more free radical damage and eventually the death of the cell. Lack of cell division also means that neurons cannot rid themselves of potentially harmful genetic mutations. Consequently, they retain damaged DNA which is of little use when the cell is under further free radical attack, leading to the vicious circle of cell dysfunction and death, especially when the cell is under stress from other sources. Old people lucky enough to inherit a robust set of mitochondrial genes may well be predisposed to enjoy a healthy old age.

The healthy brain

Before we discuss the effect on the brain of ageing damage, we need to look briefly at the healthy brain and how it works. The purpose of this section is to introduce some key terms and ideas; it can be skipped over and used as a reference section for later topics if preferred.

The brain is the most complex structure in the known universe. Weighing in at about 1.5 kilograms (3 lb), it comprises about 100 billion brain cells, each making on average about 10,000 connections with other brain cells. The size and shape of our brain structures at birth are determined by genetic factors inherited from our parents. However, as adults the fine structure of our brain is partly genetically determined and partly the product of our experiences. Although the number of genes we possess (currently estimated at 70,000) seems enormous, there is insufficient capacity in our genetic library to predetermine all the fine structure of the brain.

The adult brain is roughly spherical with an outside surface (cortex) resembling a walnut, being made up of wrinkled ridges (gyri) with folds in between them (sulci). There are two types of brain cell, neurons and glia, with at least fifty different subtypes. This enormous variation in brain cell type is a key feature of the developed brain. Neurons are complex structures which achieve a high level of specialisation within the lifetime of an individual cell; thereafter, most neurons never divide again. They are then said to be 'terminally differentiated'. This principle was first recognised by the pioneering Spanish brain scientist Santiago Ramon y Cajal. In 1913, summing up years of careful observation, he declared: 'In adult centres the nerve paths are something fixed, ended, immutable. Everything may die, nothing may be regenerated.'

As students, doctors continue to learn that the brain has no capacity to replace neurons lost through injury or ageing. Terminal differentiation gives each neuron its special characteristics. It is linked to the neuron's specific place in the brain and its complex array of connections with other brain cells. The connections are made by processes that spread out like the branches of a tree from the brain cell body. These branches, called dendrites, add up to many kilometres of 'cabling', all contained within a single brain. The cables carry electrical messages, and the longest branches (axons) – just like electrical cables – are insulated by a white, fatty material known as 'myelin'. This explains one of the simplest distinctions between brain areas. The myelin on the sheaths of bundles of axons appears white to the naked eye ('white matter'); this contrasts with areas of the brain where cell bodies predominate and there is less 'cabling' ('grey matter'). Grey matter is mostly found in the cerebral cortex and in some of the brain structures deep below the surface of the cortex.

Why the brain is not a computer

With all this 'electrical cabling', the brain is often likened to a computer. But the brain is definitely not a computer. The construction of a computer is permanent and its hardware is quite separate from the software. In the brain, in contrast, networks of brain cells organise themselves to do specific tasks. They can, in response to demands, change their patterns of connections. Experience and learning can alter the strength of these connections, their number and shape. Single networks make tens of thousands of connections that are linked to the task performed. During life, brain cells connect themselves to other cells, replace redundant connections, and build new circuits to support new tasks. The ability of brain cells to do this is sometimes called the 'self-organising principle'. No existing computer has such a capability, or approaches anywhere near the scale of structural and dynamic complexity found in the human brain.

Brain cells: neurons and glia

Neurons do most of the brain's work. Glial cells do not conduct or process information, but support the work of the neurons, maintaining and repairing them (a process sometimes called 'housekeeping'), and are also important in early brain development. Many clear and very readable accounts of the brain structure and function are available. Perhaps the best presented is *Essentials of Neural Science and Behaviour* by Eric Kandel, James Schwartz and Tom Jessell.

Neurons are found largely in the cortex, where they commonly have numerous branches arising from their cell bodies. The nucleus of the neuron is surrounded by a complete membrane to separate it from the cytoplasm of the cell. Within the cytoplasm there is another membrane folded and studded with granules (ribosomes). This is the granular endoplasmic reticulum, and is the primary site for the neuron to make new proteins. The cytoplasm also contains a single large structure of smooth membrane called the Golgi apparatus. There are many other smaller spaces in the cytoplasm and each is surrounded by a membrane: they are named according to their place, shape or size – 'vesicles' at the exterior boundary of the cell, while towards the middle of the cell are 'tubules', 'saccules' and 'vacuoles'. As in other cells, the cytoplasm contains the mitochondria, where respiration takes place, releasing energy from glucose.

Microtubular proteins

The neuron also contains three types of proteins plaited into fibrils (fibrillary proteins): microtubules, neurotubules and microfilaments. Microtubules are unbranched tubes about 25 nanometres in diameter (a nanometre is a millionth of a millimetre) with a pale centre or core about 15 nm in diameter, and appear to be haphazardly distributed throughout the cytoplasm. The main component of microtubules is a protein called tubulin. Its function is suggested by its similarity to another protein called actin, which gives muscles the capacity to contract. Microtubules form the internal skeleton of the neuron (cytoskeleton) – an internal arrangement of tubes which helps to keep individual cells in their characteristic shapes; they also form part of internal transport system inside the neuron. Neurotubules are smaller and dispersed as tiny spiral protein threads with an internal diameter of about 10 nm. Chemically, neurotubules and microtubules are quite different. Microfilaments are found in most types of neuron, often close to the cell surface when they appear as protein strands about 6–9 nm long.

Synapses are neuron–neuron connections

The surface membrane surrounding the neuron, as in other cells, is a fluid structure made up of two sheets of large molecules, and is studded with large cell surface protein molecules. The size and complexity of these proteins is the product of over 500 million years of evolution. Some are intimately involved with the regulation of information processing by the neuron; some are cell surface recognition markers; and others are concerned only with communication between brain cells. As a group they are termed 'large regulatory biomolecules'. Because they are large and complex they demand a great deal of care to make and keep functional. They are essential to many body processes, and when they eventually wear out they must be replaced. When their replacement is being constructed errors may creep into the stored design (DNA) or manufacturing process. Wear and tear on large biomolecules occurs with age, and the chances are increased that the DNA design may have been corrupted, for example, by harmful chemicals from the environment.

Brain cells talk to each other all the time. Much of this is fairly low-level chatter. Their talk is almost entirely through structures called synapses. These are sites of adhesion between neurons and are very difficult to break without drastic treatment. The neurons are separated at the synapse by a gap of about 10–20 nm (the synaptic gap). The synapse allows signals to pass from one neuron (the presynaptic neuron) to another (the post-synaptic neuron) in one direction only. In the cytoplasm of the presynaptic neuron there is a concentrated array of vesicles. These contain the neurotransmitters, molecules that are released into the synaptic gap where they attach themselves to proteins on the post-synaptic membrane. Here, there are large membrane-spanning proteins (receptors) each of which recognises just one type of transmitter. The receptor protein molecule can then change shape to allow electrolytes (dissolved salts) from the extracellular fluid to move into the post-synaptic cell and start a cascade of molecular events inside the neuron.

Neurotransmitters released into the synaptic cleft are quickly destroyed by enzymes. In brain ageing, and especially in Alzheimer's disease, the neurotransmitter acetylcholine is reduced because brain cells using acetylcholine begin to die.

Some receptors are positioned away from synpases; these can combine with molecules called 'neurotrophic factors' which assist brain development by promoting specialisation of selected developing neurons (a process called 'neuronal differentiation'). They later play a major part in helping neurons to maintain themselves and repair damage (housekeeping).

Synapses almost always occur between a long, thin extension of one neuron (the axon) and a point on the spiny dendritic sprouts of another neuron. Although synapses can occur between axons and cell bodies, this pattern is uncommon. Synaptic connections show up characteristic patterns in various parts of the brain and the richness of the branching (dendritic arborisation) with accompanying dendritic spines is an important feature of the fully developed adult human brain. In fact, loss of dendritic arborisation, with accompanying reduction in the number of dendrites, is a feature associated with age-related memory loss and dementia.

Fine brain structure

The brain's synaptic patterns are not static. There are some fixed ('hard-wired') connections between neurons, but most are changing constantly. The health of a neuron depends in part on the stimulation provided through the connections it makes with other neurons. It also depends on the supply of neurotrophic factors from adjacent neurons and the glial cells.

Diet is crucial to the developing brain. The fat soluble vitamins – especially vitamin B_{12} and folate – are crucial to normal brain development. Fatty acids from the diet are the building blocks of the brain, and a diet deficient in one of the essential fatty acids may hamper brain development. This is believed to account for the possible advantage that breast-fed children possess over bottle-fed infants in their rates of cognitive development. When certain enzymes that process fats in the brain are ineffective, brain development can be so abnormal that a severe type of learning difficulty develops from early infancy. As we shall see later, an important genetic susceptibility factor for Alzheimer's disease is closely involved with the transport of fat into the brain: slight differences in the structure of this fat transport protein (apolipoprotein E) produce major differences in the risk of developing Alzheimer's disease in late life.

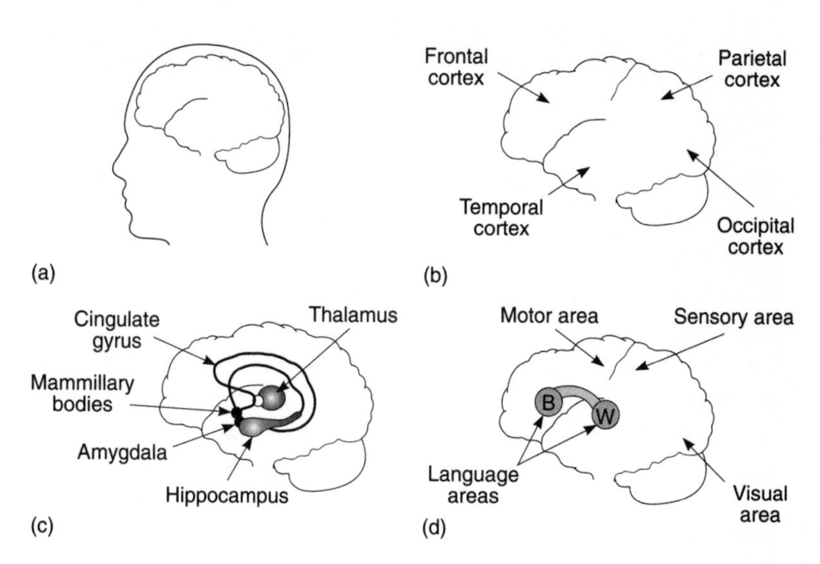

Figure 3 **The cerebral cortex** is the outer part of the brain. Figure 3 (a) shows the brain inside the cavity of the skull; the cortex is all that can be seen. The main cortical regions are identified in 3 (b) which names the main areas of the cerebral cortex. Figure 3(c) shows the limbic system deep inside the brain and important in memory and emotions. 3(d) shows how functions can be linked to areas of the cortex. (B) is Broca's area and (W) is named after Wernicke – two areas which are important in language.

The dendrites are the most striking feature of the neurons. The richness of their branching patterns and the number of spines on each dendrite are probable indicators of the pattern of synaptic connections: in some brain areas, for example, axons may run at right angles across a dendritic field with each dendrite making a single contact with the axon. The cerebral cortex comprises layers of different cell types that vary in thickness and density in different parts of the cortex. Synaptic patterns also vary between cortical areas, probably reflecting differences in function. Figure 3 shows the distribution of the primary cortical areas. The visual cortex (at the back of the brain) is the best-understood part of the brain in this regard. It has cell types that respond to different types of movement (for example, horizontal or vertical). The detection of the movement may depend on the orientation of the axon carrying the input signal to the dedicated field of the visual cortex.

How the brain makes its cortex (corticogenesis)

Almost forty years ago David Hubel and Torsten Wiesel were awarded the Nobel prize for their work on the visual cortex. They showed that this part of the brain's cortex is subdivided into a large number of functional columns. Although the overall plan of synaptic connections in these columns is determined at first by genetic factors ('nature') the subsequent fine detail of the connections is determined by visual experience ('nurture'). The visual cortex is connected to the retina, the sheet of light-sensitive cells at the back of the eye. The brain cells of the visual cortex are linked to specific cells in the retina. More sensitive parts of the retina are connected to a greater number of adjacent brain cells in the visual cortex than are less sensitive parts.

The visual functional columns of the cortex do not simply respond to light or dark but have exquisite responses to the orientation of images on the retina. During a sensitive or 'critical' period in brain development in early life, the orientation-detector cells of the visual cortex are in an unusually 'plastic' condition. At this time they can be 'sculpted' by the environment in a way not possible in the adult brain. This type of 'brain sculpting' depends upon stimulation from the environment. Brain cells in the visual cortex that are not stimulated fail to form stable and lasting functional synapses. In an important sense, this principle so elegantly described by Hubel and Wiesel reinforces what we already know. It is essential for brain cells to talk to each other if they are to develop properly and survive.

Careful studies on the way brain cells connect with one another show that these connections are highly specific. They are certainly not haphazard. In brain development, the timing of these connections is probably critical. For instance, if one brain cell sprouts an axon to connect with a distant group of neurons, and that distant group is undeveloped, or already connected or committed to another group of neurons, then the planned or 'programmed' connections will not be made. Other factors are also important: these include competition between axons, the ability of an axon to recognise a 'target' field, and the eventual contribution of other synapses to the health of the neuron.

This complex organisation of the brain begins before birth, continues in childhood and is structurally mature by late adolescence. Critical periods in the development of complex systems ensure that crucial developmental tasks are completed before later, more complex, layers of brain organisation are added.

Can the basic cellular mechanisms that ensure that the cortex develops successfully in the first place be harnessed in old age to slow or perhaps prevent dementia? Certainly, if we are ever to reconstruct a human cerebral cortex damaged by neurodegenerative disease we will need to understand the molecular, cellular and functional properties of the healthy cortex. Developmental neurobiological research may eventually illuminate the abnormalities of mind that lead to the severe mental illnesses of early life (such as autism and schizophrenia) and the crippling degeneration of the cerebral cortex in dementia.

The blueprint for the human cortex is provided by genetic information from our parents. Early in the pregnancy, the unborn child experiences a rapid burst in growth of those cells destined to form the nervous system; this is the proliferative phase. Following the establishment of cell numbers, cells set out on an orderly migration to their final positions in the cerebral cortex, where they will stop dividing: first they detach themselves from their relatives, then they send out a long 'leading process' which they use to move along specific pathways. The final position of the neuron in the cortex coincides with the cessation of cell division. This location determines its functional relationships with other neurons.

The process of cortical cell migration is now being elucidated using molecular techniques. Imagine the problem of a single neuron born close to the base of the brain and attracted to a distant 'address' somewhere out in the cortex where it will be connected to the limbic system. The problem could be scaled up to that of a small child seated on the edge of a fountain in a very crowded Trafalgar Square, who is then instructed to go to a specific house next to the Thames in Chelsea. In order to do this the migrating cell must wriggle like an amoeba through the crowds of other developing neurons. It will interact with cells it meets along the way. Some will be friendly; some will not.

Cell-to-cell interaction relies on the recognition of cell surface marker molecules. Whole movement of the migrating neuron is possible because the internal microtubular structure contains proteins that can contract. Glial cells set out first on this journey and provide with their long radial processes the 'scaffolding' along which neurons can then crawl. Some neurons prefer the company of other neurons as they slither to their final destination. Movement can be regulated because contraction of the internal proteins is controlled by the flow of salts (especially calcium) into the migrating cell. Its cell surface contains pores which can be opened or closed by chemical messages from neighbouring cells. These pores are highly specific and allow only one type of salt to pass into the cell.

Recently, Wei Wu at the Laboratory of Molecular Neurobiology in Shanghai and his colleagues in the Washington School of Medicine, Missouri, reported their detection of specific molecules that guide developing axons and also discourage neurons from invading their territory. A little like American Indians might channel settlers' wagon trains into tightly confined trails to their destination – welcome as long as you don't settle here! – these molecules provide a concentration gradient along which the neuron can move. Of course, this depends on the cell being able to 'read' that gradient, and only cells with the correct surface receptor can do this.

Chapter 2

The brain formerly known as you

Physical changes in the ageing brain

Young people – at the height of their mental powers – rarely think about the effects of ageing on the brain. As we grow older, occasional mental slips become a little too frequent. Does this mean that dementia, like a thief in the night, is creeping out of some crevice of the mind to rob us of memory, understanding and eventually dignity itself? Or is this just 'old age'? To discuss these questions, we need to link the biological theme of the previous chapter with a consideration of the social and psychological aspects of ageing.

Ageing and synapses

In the last chapter, we saw how brain cells – the neurons – make complex patterns of connections with each other. These 'neural networks' enable the brain to perform essential functions such as memory or language. These networks are not fixed; on the contrary, they can adapt to experience of the surrounding world by changing their patterns of synaptic connections. This ability to change is known as 'synaptic plasticity', and its relevance to brain ageing is a topic we shall return to again in this book. It is one of the three brain features – synaptic plasticity, myelin sheaths and large regulatory molecules – that are particularly susceptible to the effects of brain ageing.

Mental slowing is the most frequently detected age-related change in brain function. Studies of simple reaction time (the time taken to respond to a stimulus) show that it slows from early adulthood onwards. Most of this slowing occurs in the central rather than the peripheral nervous system, and it is more noticeable when more complex mental tasks are performed. As yet we have no way of studying the likely link between synaptic plasticity and slowing in reaction times. Degradation of the myelin sheath, however, is an easy change to observe; it is probably caused by free radical oxidation of the fatty acids in the myelin (membrane lipid peroxidation), and probably occurs in all brains from middle age onwards. Most observers agree that since both mental slowing and the structural changes in myelin are universal in ageing they are probably linked, but this association has yet to be proved.

Electrical activity in the ageing brain

The brain seethes with electrical activity. We can detect this by attaching electrical contacts to the scalp and measuring the voltage differences between pairs of contacts, which is taken to indicate the activity of the underlying cortex. Electrical activity detected in this way is the basis of the electroencephalogram (EEG). The contacts are spaced at regular preset intervals across the scalp and electrical activity is graphically recorded. The resulting EEG record will show fluctuations in the voltage differences between contacts (the frequency), and differences in the electrical energy are represented by the amplitude of the wave. Ageing alters the typical EEG patterns. Although the frequencies change little (if at all), the overall energy is reduced by about 25 per cent. Electroencephalographic abnormalities are seen in about a quarter of old people with no evidence of brain disease and are most often attributed to alterations in the blood supply to the brain. About half of all patients with dementia have EEG abnormalities; these are usually explained by uneven loss of brain cells in the cortex.

Understanding what the EEG is telling us about brain function is one of the great challenges of clinical neuroscience. It is probably safe to reason that when a part of the cortex is used by the brain to do a special job like the interpretation of sound or visual images then the EEG record made over that area has a lot to do with processing that sensory information. So if the EEG is restricted to just one area of the brain and a regular stimulus (such as a repeated sound) is applied to the subject's ears in a soundproofed room, the EEG will record electrical events associated with processing that auditory information. This is the basis of the test known as the averaged evoked response (see Chapter 4). Changes in this response with ageing will be linked to brain changes in the efficiency with which all sensory information is processed.

Chemistry of the ageing brain

When we reach the age of about fifty, our brain begins to shrink. At this age the average brain weighs about 1.4 kg; fifteen years later it will weigh about 1.2 kg. Much of this weight loss is caused by brain cells shrinking as they lose their water content. Water loss can be detected by modern brain imaging techniques, such as computer-assisted tomography (CAT) and magnetic resonance imaging (MRI). Figure 4 shows MRI scans from two men, aged seventy-eight, one of whom is in apparently good physical and mental health. One has Alzheimer's disease: his MRI scan shows shrinkage of brain parts. The other shows changes typical of ageing. The message from the scans is clear. Our brains shrink with age, the gaps between the folds of cortex (the sulci) widen and the large spaces (ventricles) inside the brain enlarge.

The work done by brain cells also decreases with age. Even though the brain makes up only about 2 per cent of the body weight of a 70 kg man, it consumes 20 per cent of all the energy. This high consumption is maintained by the blood flow through the brain being protected at a high (and privileged) rate compared with other organs. The energy used is derived from glucose in the blood, and changes in its consumption can be measured using the method of positron emission tomography (PET). This technique relies on the detection of a special type of radiation (positrons) released from a short-lived radioactive atom by an array of radiation detectors placed around the subject's head. The procedure involves the rapid synthesis of marker molecules of glucose containing the radioactive atom ('labelled glucose'), injection of the radioactive glucose into an arm vein, followed by scanning of the head to map the distribution of the glucose in the brain. Rapid repetition allows differences between brain

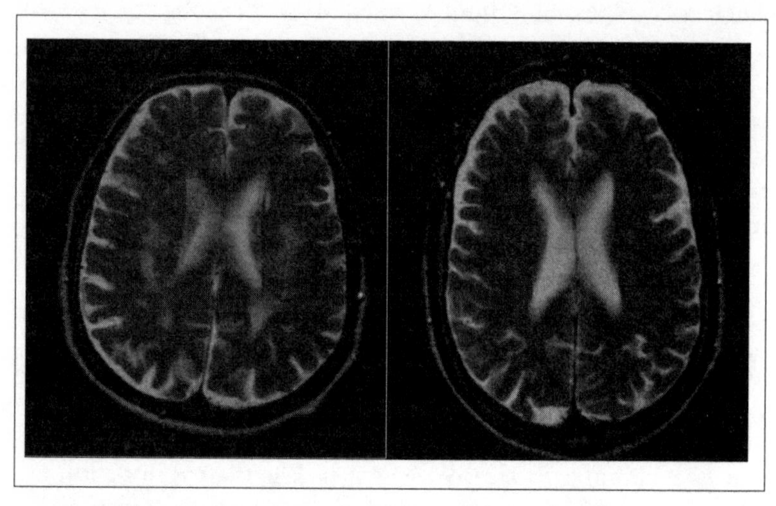

Slight cortical shrinkage Marked cortical shrinkage

Figure 4 **Magnetic Resonance Images (MRI) of the brain**. The MRI scanner reveals the distribution of water within the cavity of the skull by subjecting water molecules to strong magnetic fields and causing water to resonate and release radio energy. This radiation can be detected around the head and the sources reconstructed in a computer. Here two MRI scans from men both aged seventy-eight show how clearly MRI detects cortical shrinkage with age. The lighter shaded area in the centre of each scan is larger in the man with more cortical shrinkage. The lighter appearance is caused by the fluid content of the space releasing more radiation than the tissue of the brain. Within the tissue the MRI distinguishes between grey matter (mostly brain cells) and white matter (mostly fat-covered bundles of nerve fibres).

structures to be detected and related to the type of work each part of the brain is doing at that time.

Labelled glucose accumulates in cells doing brain work. With ageing, there are consistent decreases in the amount of glucose taken up by active brain cells. Sometimes it is possible to link these decreases to the development of mental symptoms or impairments such as memory difficulties. Unfortunately, too few old people have been studied carefully with these sophisticated techniques to provide firm conclusions about the links between brain glucose uptake changes and the appearance of specific mental symptoms or disorders. Oxygen can also be labelled with a radioactive marker in the same way, and used to measure oxygen consumption by the brain. The results are very similar to those obtained with labelled glucose.

Brain cells talk to each other using chemical messengers known as neurotransmitters. There are two kinds: 'classical' transmitters, so called because they were discovered first, and neuropeptides, whose role in neurotransmission was only established much later. Neuropeptides are present in the brain at concentrations about a thousand times less than the 'classical' transmitters. So far, more than 500 different substances have been found naturally in the brain and many of these meet the criteria for a neurotransmitter. With ageing, the amounts of specific transmitters may gradually decrease. Neurotransmitters involved with the control of movement, attention, arousal, the sleep/wake cycle, eating and aggression are found in quite separate brain structures. Reductions in the amounts of certain transmitters can be correlated with the severity of symptoms caused by neuron loss from these brain structures. This principle was first established for Parkinson's disease, then for Huntington's disease and later for Alzheimer's disease. In normal ageing, the amount of the transmitter acetylcholine slowly declines in the hippocampus. When this was discovered, it was quickly recognised that experimental drugs given to block this transmitter always caused memory problems. Likewise, damage to the hippocampus will also cause profound disturbance of memory. Loss of acetylcholine-containing cells from the ageing hippocampus is one of the most likely explanations of memory deficits in otherwise healthy old people.

Large regulatory biomolecules in brain cells are also susceptible to age-related changes. The modification of these large molecules by sugars produces complex, insoluble compounds referred to as 'advanced glycation end-products', appropriately abbreviated to AGEs. These hard, yellow-brown compounds are important in Alzheimer's disease. Work on the biochemistry of AGEs is only just beginning and reveals a chemistry of considerable complexity. Oxidative damage by AGEs to the constituents of nerve cells yields products capable of starting a cascade of further damage and the formation of abnormal protein complexes in the ageing nervous system.

So far as we have mentioned 'white' and 'grey' matter, but the brain is not so dull in appearance as this may suggest. Some regions of the brain have distinctive colours (black, yellow or orange) due to naturally occurring pigments, which can accumulate in some large brain cells. For example, the substantia nigra (literally 'black substance'), a paired structure in the base of the brain (the basal ganglia), owes its colour to the pigment neuromelanin. Another pigment – lipofuscin – gives some structures a yellow colour. These pigments accumulate with ageing, but it is uncertain whether either neuromelanin or lipofuscin is involved in the ageing process, whether their presence accumulates with local brain work, or whether they are a consequence of ageing itself.

Neuritic plaques

The best-known structural changes in the ageing brain are the development of neuritic plaques and neurofibrillary tangles.

The first neuropathologists reported these structures around or close to the blood vessels of the brain and called them 'plaques'. With the advent of electron microscopy, the plaques were seen to be composed mainly of bits and pieces of broken-up neurons, mostly membranes. When brain tissue sections were stained with different dyes, neuritic plaques were found to take up the same dyes that stained plant starches, so the term 'amyloid' ('starch-like') was coined to describe the plaque's major constituent. Later biochemical examination showed that an abnormal protein was responsible for this 'amyloid' property of the plaque. The discovery of the precise chemical composition of amyloid protein was a turning point in understanding the biochemistry of Alzheimer's disease – a detective story that is told in Chapter 9. In the ageing brain, neuritic plaques are found at various stages of maturation. These range from diffuse deposits of amyloid protein to complex lesions.

Are amyloid deposits are an inevitable consequence of brain ageing? Probably not, but no one is too sure. There are too many examples in the scientific literature of old people dying after the age of ninety with no evidence of amyloid deposition in their brains.

The first comprehensive study appeared in the late 1960s. Bernard Tomlinson and his psychiatric colleagues Martin Roth and Gary Blessed in Newcastle-upon-Tyne described the brains of 400 old people. They showed that the density of neuritic plaques counted in a standard microscopic field was related closely to the person's overall mental ability before death. This was the first scientific report to make a quantitative link between a structural feature of the brain and a mental function in life. The studies of Bernard Tomlinson and Tony Corsellis are landmarks in understanding the relationships between 'normal' brain ageing and the dementias. Before this pioneering work in Newcastle and London, most doctors thought that Alzheimer's disease was a rare form of early-onset dementia. For the first six decades of the twentieth century, Alzheimer's disease was believed to be caused by 'hardening of the arteries of the brain'. Tomlinson and Corsellis showed that not only were the brain changes in early-onset Alzheimer's dementia indistinguishable from the common late 'senile' dementia, but also hardening of the brain arteries was absent from more than half of the late-onset cases.

There are now some preliminary accounts from Colin Masters in Australia and Konrad Beyreuther in Germany of aged native black Africans who show no or little evidence of amyloid deposits. Such studies rely heavily on the care taken to exclude mental impairment before death and the extent and precision of the search for amyloid. The consensus view is that limited amounts of amyloid are common in almost all old brains and that there is a subgroup of old, non-demented people who have extensive deposits of amyloid. Although these subgroups appear not to be demented, the psychological tests used may have been insensitive to more subtle types of mental impairment. Biochemical studies of amyloid deposition in ageing suggest that there are several subtypes of amyloid protein and that the mixtures of these are different in 'normal' ageing and in Alzheimer's disease.

Amyloid protein is an abnormal substance which the body's immune system readily recognises as foreign. Not surprisingly, the microglia which function as the brain's 'scavenger' cells take a keen interest in this new protein. There are grounds to believe that amyloid protein can directly stimulate microglia. Likewise, some harmful products of cell damage (caused by free radical damage) can activate microglia by binding on to cell surface receptors. Detailed biochemical analysis of markers of microglial activation provides good evidence that a very slow, low-grade inflammatory response is taking place in the ageing brain.

Neurofibrillary tangles

Neurofibrillary tangles are found within the cell bodies of neurons in both 'normal' ageing and Alzheimer's disease. When neurons die, as in Alzheimer's disease and certain related conditions, neurofibrillary tangles are found between the surviving neurons. They have been likened to the 'ghosts' or 'gravestones' of dead neurons. Neurofibrillary tangles are made up of aggregates of neurofilaments, revealed under the electron microscope to be pairs of 10 nm diameter neurofilaments arranged as a double helix.

Henry Wisniewski is a Polish doctor working in New York who perfected the use of the electron microscope to study normal brain ageing and Alzheimer's disease. His work on the formation of neurofibrillary tangles revealed their detailed structure, a possible environmental trigger (aluminium), and the finding that they occur invariably in Alzheimer's disease but are sparse in normal ageing. This work was fundamental to valuable research programmes around the world. In Germany, the husband-and-wife team of Heiko and Evo Braak examined 800 brains of demented and non-demented people aged between twenty and ninety. They found that the deposition of neurofibrillary tangles may be compatible with 'normal' ageing when this occurs in brain structures non-crucial to memory. In England, at the Laboratory of Molecular Biology in Cambridge, Michel Goedert and his team have tackled the problem of neurofibrillary tangle formation in health and disease using powerful molecular biological and molecular genetic techniques. A key constituent of the neurofibrillary tangle is derived from a naturally occurring protein (tau) which the brain cell uses to build its microtubules (see Chapter 1). This microtubular assembly protein is abnormal when neurofibrillary tangles form. It is hyperphosphorylated (i.e. it contains an additional phosphate group), making it more likely to form large aggregates or to clump.

When the deposition of neurofibrillary tangles extends to involve brain structures essential to memory, then the symptoms of Alzheimer's disease are inevitable. Although the study data are compelling, too little is known about how memory works in normal subjects for us to accept easily the idea that there is a continuum between 'normal' ageing and Alzheimer's disease.

Are brain cells lost with ageing?

The whole brain

As we age, the total volume of brain tissue slowly shrinks. Using modern brain imaging techniques, the volume of the whole brain, its white and grey matter content, and the volume of surrounding fluid (cerebrospinal fluid) can be measured. The pioneering studies of Harold Brody in the 1950s are the basis for the claim that the whole brain shows a gradual decrease in brain cell number with ageing. His work was based on comparisons of brain weight between groups of different ages. His results, however, did emphasise that some parts of the brain seemed more vulnerable than others. For example, modern techniques used by researchers following Brody's trail have shown that the brain cells concerned with processing movement instructions are more extensively lost than cells in general. These changes are found in the cerebellum and the basal ganglia. The frontal lobes of the ageing brain show more pronounced thinning of the cortex than any other part of brain. These findings are consistent whichever method of measurement is used. Magnetic resonance imaging can detect greater decreases in the volume of the frontal lobes than in any other part of the brain cortex. For example, one study of healthy old people found the annual rate of volume decrease in the frontal lobes was twice that of other brain regions – 0.55 per cent compared with 0.27 per cent elsewhere. Studies using PET scans support this observation. When blood flow in different brain regions is compared, the flow to the frontal lobes shows a much greater decline than does any other brain region.

The modern view now runs counter to the historical position. The older perspective saw normal ageing as a cascade of insults to the brain ultimately leading to the death of brain cells. Early pathologists could see no evidence that the brain could replace the lost cells by cell division; instead, they found evidence to support their idea that brain cell loss was the cause of normal brain ageing. By counting brain cells in ageing Old World primates, significant age-related decreases in the density of brain cells were observed in areas (such as the hippocampus) critical for memory. The greatest loss was always seen in older subjects and in those with the biggest memory problems. Importantly, animal studies are directly relevant to human brain ageing. As our species has evolved, so our brains have increased in relative size and become more complex. The frontal lobes of humans are much larger than those of the chimpanzee, our closest relative, and take up about a third of the volume of the entire brain cortex. They are regarded as the most recently evolved parts of the brain and are believed to be crucial for many of the intellectual functions (including planning and abstract reasoning) which rely on mental imagery, characteristic of insightful problem solving.

Brain cell loss, however, is no longer considered to be the prime cause of mental slowing and memory decline in old age. Modern studies in humans and other primates show that brain cell loss is much smaller than previously claimed. Moreover, it does not occur evenly throughout the brain: some parts are more affected than others, in a process termed 'selective abiotrophy'. Some brain diseases are very specific; populations of brain cells are selectively lost in Alzheimer's disease, Parkinson's disease and motor neuron disease. Many brain scientists have now shown independently, using sophisticated modern techniques, that there is little evidence for the old-fashioned view that extensive brain cell death occurs in 'normal' ageing. This is as true of old people as it is of old monkeys and old rats.

Selective cell loss in key brain structures

The hippocampus is crucial to memory function. The number of brain cells in the hippocampus is preserved in old people, even in the presence of memory problems before death. This observation demolishes the earlier theory that brain cell loss in the hippocampus and its subsequent shrinkage were the principal cause of memory impairment in normal old age. Now we know this is not true. The hippocampus only loses brain cells as part of a disease process. Cell loss is not a prerequisite for age-related decline in learning or memory.

The frontal lobes of the brain cortex, as we saw earlier, are especially affected by ageing. There is now a consensus that the psychological changes (mostly deterioration) of ageing reflect deterioration of the frontal lobes of the brain. Louise Phillips and Sergio Della Sala in Aberdeen have proposed a specific neuranatomical theory of cognitive ageing, based on the usual pattern of impaired and spared mental functions found typically in ageing. They noted the profound similarities (and the exceptions too) between 'healthy' ageing and damage during adult life to the frontal lobes of the brain. These ideas are influential in current psychological and clinical imaging studies. However, the size and complexity of the frontal lobes make simple assumptions about their precise role in ageing almost impossible to argue successfully. First, like the rest of the brain cortex, the frontal lobes are not symmetrical, one side being 'dominant' over the other. Second, there are distinct anatomical areas in the frontal lobes, which are not equally affected by ageing processes. This is particularly true of comparisons between the dorsolateral and orbitoventral areas of the frontal cortex. Once again, comparisons between humans and primates tell us much about the differences between these areas. The dorsolateral frontal cortex was the last part to evolve in brain development. This observation encourages some to believe that the most recently evolved areas of brain cortex are those most affected by ageing.

Looking beyond the hippocampus, other parts of the brain cortex are also spared in 'normal' ageing. This is true even for parts of the cortex that support types of memory typically impaired in 'normal' ageing. Outside the frontal lobes, the cortex is relatively spared; but is the same true of structures that lie deep below the cortex? The most extensively studied subcortical system is the group of brain cells that use acetylcholine as a neurotransmitter and connect with the frontal cortex.

In 'normal' old age, these brain cells are selectively lost in humans, monkeys and rats, irrespective of the part of the cortex onto which these acetylcholine-using (cholinergic) cells project. Careful observations of acetylcholine content show that the extent of loss of acetylcholine is closely related to the decline of memory function in 'normal' old people. However, decreasing acetylcholine transmission does not explain everything about loss of cognitive functions in old age. Many other chemical neurotransmitters and neuropeptides are also involved. Countering the historical view, a new concept is emerging as a result of detailed studies of the ageing brain. The cortex is not particularly vulnerable to ageing, but its subcortical structures seem to be so. Particular transmitter systems are more affected than others. The reasons why this is so, and why particular cell types seem more susceptible to ageing, are among the major questions in understanding the causes and effects of normal ageing.

Synaptic plasticity and successful brain ageing

Although many of the original patterns of connections between brain cells established during development remain fixed, reinforced by continuous use, many others are in a state of flux, making and keeping contact, withering with disuse, or changing properties. Experiences are important in changing these patterns, which in turn seem likely to be involved in the formation of certain types of long-term memory.

Robert Terry was one of the first neuropathologists to use the electron microscope to compare normal brain ageing with Alzheimer's disease. He started out at the Albert Einstein School of Medicine in New York and moved later to the University of California in San Diego. He made one of the major discoveries in Alzheimer's disease when he showed an even closer relationship between brain structural change and dementia severity than that reported for neuritic plaques by Bernard Tomlinson in Newcastle. Terry's technique relied upon simultaneous improvements in the use of computers to image and measure brain cell structures and the refinement of techniques from immunology to characterise the components of brain cells. His work on synapses – the point of information transmission between brain cells – showed that the decrease in number of synapses in the brains of patients with Alzheimer's disease could not be explained by a reduction in brain cell numbers. Decreased synpatic density proved to be closely linked to the extent of mental impairment before death in Alzheimer's disease. At this stage, it was reasonable to propose that decreased synpatic connections probably represented failure of mechanisms underpinning synaptic plasticity. This failure could be linked (speculatively at the time) to memory impairment in the old, and perhaps even to dementia.

In the face of the selective loss with ageing of neurons from specific areas, the brain succeeds in preserving its essential circuitry. In part this reflects the adaptive capacity of the brain (a feature of its 'self-organising principle', discussed in a later chapter), and also the fact that the brain retains some developmental capacity throughout life. The capacity of the brain to adapt to loss of brain cells has a biological basis and is now fairly well understood. As brain cells die, their processes wither. Cells that once received information from those now dead or dying brain cells respond by releasing chemicals that stimulate surviving brain cells to make new connections. The American researchers Buell and Coleman in New York examined these processes in normal ageing and in Alzheimer's disease; they found that in normal old people the branching of processes away from brain cells is more extensive than in middle-aged adults. Unfortunately, these new outgrowths do not occur throughout old age. After the age of about seventy, the capacity of surviving brain cells to respond to brain cell loss in this way is gradually diminished.

Chapter 3

Any cook could run the country

Social and psychological aspects of ageing

Ageing has had an enormous impact on modern society. Old people are now the fastest-growing segment of the population of most developed countries, and their increased expectations of health, social involvement and achievement challenge the dogma of inevitable age-related decline. The greater expectations of those approaching old age rest on the evidence of their eyes; they can see that old people now live longer and healthier lives. Previous generations suffered greatly from the indignities of disability and dependence in old age; now there are substantial numbers of healthy old people who are active and independent. These improvements in the health of many old people are often attributed to social policies related to health care, housing, and better opportunities to maintain social relations.

The social dimension

Two social factors have emerged as powerful predictors of the risk of age-related mental decline and poor quality of life: these are lifelong inferior general health, and low educational levels. Obviously, neither is a primary cause of brain ageing. A moment's reflection also shows that these factors are not independent of each other. Poverty, poor education and diet, chronic ill-health and stress are all-too-familiar elements of a disadvantaged life, and any one of these elements could be working as the predictor of brain ageing.

Social ageing raises the most complex issues of all – not least because no one is sure of the extent to which social ageing causes mental decline, or whether mental decline causes social ageing. Early laboratory experiments showed that an enriched social environment improves overall mental performance. This has been shown to be true for patients recovering from stroke as well as in animal studies. Community-based studies show time and again that a rich social network has positive benefits for health and for the integrity of body systems. Not surprisingly, these benefits extend to mental health and include the preservation of mental ability.

Many other social factors contribute to the quality of life in old age. Family and schooling, attitudes towards ageing, and social involvement seem to be important. Maintenance of mental effort, certain features of character and temperament, and interest in the meaning of personal experiences all define an individual's self-perception and influence his or her choice of social contacts. Although each of these features influences changes in social adjustment with ageing, none provides a basis for a totally satisfactory social definition of ageing. As with biological and psychological definitions, the best solution may be to ignore the impulse to define, and instead to examine the underlying processes that change with time.

Social behaviour changes with ageing, though not in any truly consistent fashion. A great deal is made of the process of social disengagement whereby old people gradually withdraw from society. Friends die, family members move away, journeys become much more effortful and are overshadowed by a fear of falling. Disengagement seems a rational and necessary piece of personal economy; unfortunately the consequences for mental health may be disastrous. Without a social network to provide advice, support and comfort, molehills become mountains, strangers become potential assailants, and lapses of memory are erroneously magnified into convincing signs of imminent dementia. Depressive illnesses are the best-known and are sadly the most frequent unwanted consequences of social withdrawal. They cause much unnecessary suffering, are potentially preventable, and are often readily reversed with treatment.

What is an old person?

If we cannot define social ageing adequately, then can we define what an old person is? This is just as problematic. Because of the enormous individual differences between ageing people, a simple calendar definition of an 'old' person is hard to defend. Many people regard the state age of retirement as the watershed between adulthood and old age, but this is largely an arbitrary division. In his *History of Old Age*, George Minois points out that from antiquity to the industrial age, retirement at a particular age was an option open only to affluent Europeans. Everyone else worked until they were no longer able to do so. The retirement age of sixty-five years is usually attributed to Germany's Chancellor Bismarck who, when pressed to pay pensions to war veterans, reluctantly agreed. His advisers then queried when these veterans might become eligible for their pension. 'How old are they when they die?' Bismarck is reputed to have demanded. 'Around sixty-six,' came the reply. 'Then they get the pension at sixty-five,' he retorted. The entirely arbitrary age distinction between old and young is further exemplified by a distinguished professor who devoted his life to the study of dementia. In his early writings he divided dementia into a 'presenile' type (before age sixty-five) and a 'senile' type (after that age). Later on, he decided that 'early' dementias occurred up to the age of seventy, and in even later publications seventy-five became the dividing line. As he passed his own retirement age and continued his research, he chose to divide the two types of dementia at age eighty. Like beauty, the question of who is an old person can lie in the eye of the beholder.

So old people do not make up a homogeneous group. As with any other developmental epoch, individuals enter 'old age' with their own resources, strengths and vulnerabilities. Failures to accomplish the maturational tasks appropriate to earlier stages of development can become handicaps and increase the likelihood of unsuccessful ageing. Damaging experiences in adulthood (for example head injury or severe depressive illness) may be linked to an increased risk of dementia ten or even twenty years later.

For many reasons, old people cannot be casually compared across generations. When we reach the age of eighty, our brains may not show the same age-related changes of 'normal ageing' as did those of our parents and grandparents when they were that age. They might have lived through the poverty of the depression in the 1930s or the food rationing of the Second World War. The post-war generations were less often breast-fed and generally had a richer diet than their forebears. Younger generations, especially in the developed world, have been experimenting extensively with psychoactive drugs, and we can only be uneasy about other effects of the modern environment (diet, alcohol, microwaves, and so on) on the processes of brain ageing. These issues show that it is unsafe to draw too many conclusions about ageing if our data are obtained from comparisons between individuals of different ages. Studies of the same individuals over the entire life span are the best source of information.

This caveat is interesting for another reason. There is evidence of a greater than expected number of 'old' people born between 1910 and 1925 whose body functions show only minimal decline in performance compared with previous generations. Although this subgroup is fairly small in the developed world, there are anecdotal accounts of healthy, extremely long-lived people in geographically isolated populations in central Asia and South America. Some researchers believe that the isolation of these often illiterate societies leads to exaggerated estimates of the life spans of their very old members; other researchers suspect that their longevity may be due to their different lifestyles.

The role of disease

Many common disorders of old age do not affect the brain directly but do impair brain function, and must be taken into account in studies of brain ageing. What are the common disorders and diseases that affect brain function in old people? Are they sufficient to cause the symptoms of brain disease? Are the changes in the structure of the brain that occur in 'normal' ageing similar to – albeit less severe than – those of Alzheimer's disease? These questions touch on the important idea that a certain 'threshold' of structural change in the brain must be exceeded if symptoms are to appear. This 'threshold' has many determinants, and, over the life span, can be influenced by both genetic and environmental factors. These issues are examined in later chapters.

The psychological dimension

The preceding discussion on social aspects of ageing looked at the interaction between the ageing brain and society. That this is a two-way process may not have been immediately obvious. We shall now look at some of the psychological changes that occur with ageing: here the two-way link will be clearer. There is no exact division between social behaviour and the realm of psychology: indeed, there is even a science of 'social psychology'. For convenience, the focus of this section is on those mental abilities that change consistently with age.

The most obvious psychological impact of ageing occurs on retirement from work. This is not a sudden cessation. The productivity of most workers declines from the age of about forty. Diminished energy and motivation, growing non-work interests, and the experience of illness (whether personal or in someone close) compose a well-known set of factors that reduce work output. These changes are of such economic importance that reports abound of the nature, causes and consequences of the effects of ageing on work performance.

The age-related decline in mental speed and intelligence from early adulthood (around age twenty-five) may partly explain the decrease in work efficiency. However, decline in these mental functions is more than compensated for by the acquisition and consolidation of other mental skills and problem-solving abilities, which should make older employees more valued, not less. Paul Baltes of the Max Planck Institute in Berlin has written extensively about the psychology of ageing. He has good evidence to show that while some abilities decline, others can improve, especially if training is given.

Psychological ageing probably begins around the age of fifty and is well established by the age of seventy-five. It is not a simple matter of mental decline and senility. Many old people show a decline in some performance abilities but a gain in others. For example, they may come up with fewer solutions to everyday problems, but these can more often than not be the solutions that work best for them. The idea that changes in mental abilities need not all be harmful for the individual is gaining ground among psychologists. The fact that some old people actually improve at some types of mental performance task is now taken seriously in scientific studies, and it is possible that these types of improvement have biological components open to analysis.

It is a common observation that one person is 'old' at sixty but another is as sharp as ever at seventy-five. In terms of mental abilities, long-term studies show consistently that some people maintain adult competencies into advanced old age (at ninety and beyond) whereas others decline. These losses occur in mental speed of response, the retrieval of recent information from memory, and the use of mental imagery to find their way around (spatial ability). Findings of this type were first reported from what has become the best-known study of psychological ageing. This began in 1956 in the University of Washington, Seattle, as the basis for the doctoral thesis of Warner Schaie. Before we look at Schaie's work, a simple metaphor is used to introduce some terms and ideas from psychology.

A metaphor for the mind

In general terms, you do not need to know much about theories of thinking to understand modern research on the ageing of mental abilities. You can survive on the meagre rations provided by simple descriptions of the main changes in intelligence, memory and language. But it will not be a diet to promote intellectual growth. You are unlikely to ask the right questions without some grasp of the science of 'cognitive psychology'. This science has given us a broad classification of the mental phenomena that may change with age: attention, sensory systems, thinking and imagery, strategies and planning, emotions, memory and language.

If you remained at this level of simple classification of phenomena, the approach would be of limited value as it only sets out broad categories of mental change. At a slightly higher level, cognitive psychology (literally the psychology of knowing) offers well-tried and trustworthy methods of scientific enquiry, suitable (for example) for comparing short-term and long-term memory differences between young and old subjects. Go higher still, and cognitive psychology starts to assemble a theoretical framework of mental life, combining what is already known with sensible predictions about what needs to be known. This integrative approach will generate new ideas for research,

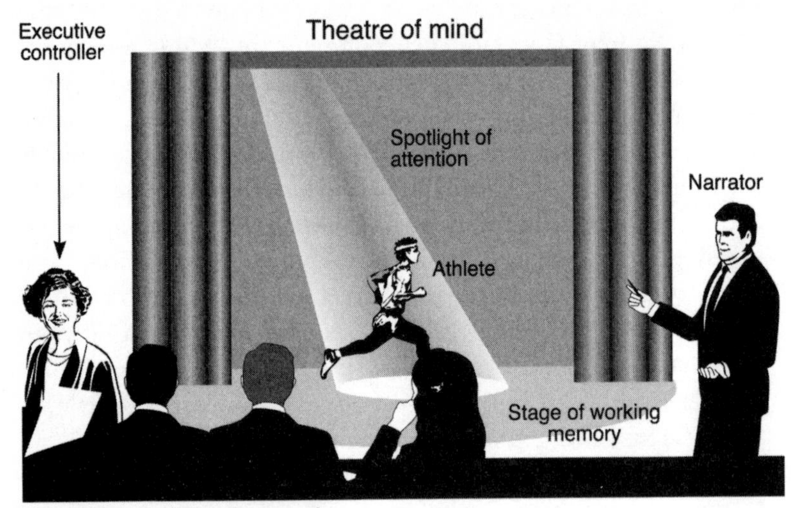

Figure 5 **The Theatre Metaphor of Mind.** A spotlight (attention) illuminates part of a stage onto which a runner has appeared. A narrator stands to the right and maintains continuity between scenes as they take place on stage. An executive controller stands to the left and directs the involvement of specific parts of the audience – brain modules – that actively perceive what the senses provide. These modules and attention are under the direction of the executive controller.

improve understanding of brain ageing, and eventually provide guidelines for the treatment and prevention of the loss of mental abilities with ageing.

Cognitive psychology is too large a topic to cover fully here. Instead, let us look at one particular model of cognitive psychology, which uses probably the simplest of metaphors for the workings of the mind: it is based on the stage of a theatre, where actors play parts before an audience.

The 'theatre' metaphor of the mind has been around since the time of Aristotle. It uses the idea that mental processes can construct and view an internal stage on which are displayed representations of the outside world. The theatre 'audience' is made up of parts of the brain that must process and make sense of what is 'on stage' at any moment in time. These brain areas that make up the audience include arrays of memory banks. Parts of the audience may be directly linked to consciousness, others comprise an 'unconscious audience framework'. Figure 5 shows this metaphor in cartoon form. Note the role played by 'attention', shown as the theatre spotlight. This is under the direction of a central 'executive controller'. There is also, rather like a medieval mystery play, a 'narrator' describing what is happening on stage. Unlike scripted theatre, this narrator cannot be sure what will happen next, and must try to make sense of it all. Gaps are filled in by the memory banks in the 'unconscious audience' that come up with the best-fitting bits of continuity material. This simple metaphor can be used to illustrate the effects of ageing. These concepts are not without difficulty; even experienced researchers find brain studies sometimes hard to understand. But then, as patients also say, if our minds were so simple we could easily understand them, then *we* would be so simple we could not understand them.

Ageing abilities

With these thoughts in mind, what predictions can we make? First, we could predict that ageing is generally bad for the brain. According to this idea, anything we can think of that depends on the brain to function efficiently is slowed down and made inefficient by ageing. Over the past twenty years, this has been one of the most influential theories of ageing. Sometimes called the 'information processing' theory of ageing, it links general mental slowing with decreases in performance at all mental tasks with ageing. This theory now has to vie with newer, more elegant models, which propose that age-related deterioration in some – but not all – brain regions with ageing disrupts some mental functions while sparing others. These patterns of deterioration also disrupt connections between brain regions, and the resulting impairments of brain integrative function may cause disintegration of, for example, social judgement and emotional responsivity.

These theories are founded on a large body of observations of brain structural and psychological changes with ageing. While a good measure of agreement exists that the structure of the brain is unevenly affected by ageing, it is by no means agreed that psychological changes correspond exactly with the structural ones. This uncertainty points to an important conceptual problem in brain ageing studies concerning the correlation between brain structure and function. The pattern of structural change described earlier argues powerfully against the idea that ageing affects the brain in general or 'global' fashion. Some brain areas such as the frontal lobes age faster than others. However, these age-susceptible areas are among the most recently evolved and do not lend themselves easily to correlative structure/function studies of the kind required to map psychological performance onto the brain cortex. In the early 1980s Jerry Fodor proposed a modular theory of brain organisation, in which the modules were computational arrays of neurons that maintained a fixed pattern of connections with one another and served a specific psychological function (or 'domain'). These modules form part of the 'conscious audience' framework (Figure 5) and are capable of rapid, large-scale processing. Typically, brain scientists think in these terms of the perceptual processing of sensory information such as the component sounds of speech or visual images.

The modular theory of brain organisation does not explain the brain's powerful ability to integrate information between different domains. This ability is global, works rather slowly, and is under control of the central executive. Problem-solving using mental imagery is a good example of this type of integrative ability. Another problem with the modular theory (as with many subjects in science) is that the more closely the proposed modules in the brain cortex are examined, the less clear seem the structural boundaries between them – and the less 'modular' they appear. From an evolutionary perspective, the brain appears to have evolved its modular structures first and then acquired its cortex, which operates best when the boundaries between modules are loosened. However, it has already been pointed out that structural changes in the brain are uneven, and might be regarded as 'modular'. Is this sufficient to support a modular theory of brain ageing? If so, it would provide strong support for the idea that brain ageing is selective and not generalised.

There are problems with uncritical acceptance of this proposal. Many brain functions clearly rely on cooperation between cortical areas, so brain ageing cannot be simply modular or selective. The example of intelligence will illustrate this point. As an alternative, we could predict that ageing affects brain structures and functions in a haphazard and totally random fashion; in this second possibility, it would just be a matter of luck if something were lost or retained. A third scenario predicts that the effects of ageing are quite selective: ageing would affect specific major brain structures and these effects would be consistent in extent and timing. These structures might, for example, provide storage of long-term memories or support attention or memory retrieval. The same scenario might predict the sparing of highly practical skills. A fourth suggestion is that ageing gradually restricts the brain storage capacity. According to this hypothesis the accumulation of life's experiences slowly fills up the available storage space. In turn, this space would be gradually corrupted as an inevitable consequence of brain ageing.

These alternatives become three simple questions about brain ageing:

1. Is brain ageing general or selective?
2. Do brain ageing changes occur at random?
3. Does ageing drain the brain's overall capacity?

Mental life in old age

Some mental health symptoms become more common with age. One of these is the pervasive sense of despair which so often affects old people afflicted by loss. Bereavement may seem the most obvious form of loss, but it is by no means the most common. We can lose skills, well-being, hearing and vision. We can lose a rich and varied social life. We can lose these things slowly or quickly, with greater or lesser immediate impact, with consequences that show considerable differences between individuals, but even the most resilient will feel a sense of impoverishment by significant loss. At times we may dwell on the meaning of it all, feeling despondent and condemned to the apparently inevitable. This is the first problem in studying changes in mental ability in ageing. Some old people won't want to take part; others may try to help, but feel they can't do themselves justice. Certainly, some may be less than wholehearted in their approach to mental ability tests given as part of someone's research project on the psychology of ageing – particularly, if the tests resemble those last seen at school sixty years previously. If school was never remembered as much of a success, then failure (again) is the best that can be expected. Negative thoughts such as these are usual in depressive illnesses. They also occur frequently in people who are not clinically depressed, but who are rather better thought of as demoralised by ageing. Poor performance on a specific mental test by an unhappy old person who has experienced the deprivations of old age may not reflect a genuine reduction in a specific mental ability. The poor performance may instead be attributed to reduced motivation and to drive diminished by circumstances rather than to ageing itself.

We know that depressive illnesses are common in old people. When old people who meet strict criteria for depressive illness are given mental ability tests they typically do less well on all aspects of the tests, not just in ways that suggest selective loss of brain function. Their errors tend to be slight and are sometimes made with much self-critical comment. There are also frequent complaints of poor concentration, tiredness or physical discomfort with the test setting. Not surprisingly, the inclusion of demoralised, unhappy, depressed old people in samples of the general population may give a false impression of the reduction of mental abilities in that population. The more such people you include, the more the total population might seem to be impaired by ageing. This is an important source of error in community studies of changes in mental ability with age.

Sensory loss

Losses in sensory ability – particularly hearing – are among the most obvious mental changes of old age. From the age of about twenty-five, hearing loss can be detected in both men and women. From the age of about thirty-two in men and thirty-seven in women, people begin to complain about their hearing, and as ageing proceeds, perception of higher frequencies (usually in the 4000–6000 Hz range, but most marked at over 8000 Hz) is reduced. There is also a quickening of the rate at which a sound is perceived, so that old people complain that noises are louder than they are in reality. Hearing losses of this type are extremely common and may impair not just the appreciation of music or the spoken word, but also the capacity to take part in conversations, and to make and keep new friends.

Sight is also impaired by ageing. The principle of synaptic plasticity – how experience sculpts connections between brain cells – was first worked out for the visual cortex. With ageing, this same plasticity can compensate for visual impairment. The best-known example is the clouding of the lens of the eye that begins at around the age of twenty and becomes noticeable to the naked eye as a cataract in the lenses of some old people. This clouding of the lens reduces the amount of light falling on the retina at the back of the eye, and scatters the light that does get through. As the brain slowly ages it has plenty of time to adjust to the changed properties of the image relayed to it by the retina. One of the most remarkable experiences for an old person is to have vision once impaired by bilateral cataracts restored by replacement lenses. Artists who have undergone this operation have been amazed at the extent of their former misperceptions of the world when they compared work completed before and after the replacement.

Our eyes are able to adjust themselves to different light levels. This property of adaptation becomes less efficient with age, particularly after the age of sixty. Colour vision and changes in visual acuity also worsen with ageing: it becomes more difficult to distinguish blues from greens, and both near and distant vision are affected. The most important consequence of such visual changes in everyday life is their effect on the ability to drive a car.

In mental test settings, two other consequences of visual impairment emerge as possible sources of error. First, reading speed may be slowed, especially if the print is small. Second, visual search strategies may become less efficient. These consequences must be considered if computer screens are used to test old people.

The senses of touch, pain sensation and balance are also diminished by ageing, but are less likely to influence the outcome of most tests of mental ability in old age.

Mental life, physical health, prescribed drugs and alcohol

Decades of research on the psychological effects of drugs prescribed to reduce anxiety or induce sleep, and of non-prescribed sedative drugs such as alcohol, show how easy it is to underestimate their considerable effects on mental ability. The reasons for these harmful effects are numerous and complex, but the consequences are simple to understand. These drugs make ageing seem worse than it should be.

Drugs that impair mental function may continue to do so for many hours after ingestion. Worst affected are our powers of attention and correct orientation. Using our theatre metaphor, it is as though the drug could block instructions from the central controller to the spotlight operator. Attention and awareness are closely linked: when attention is impaired, so is awareness. *Selective attention* is the ability to throw the spotlight onto a limited range of stimuli. *Divided attention* is the ability to do several things at once; in other words, the 'executive controller' can supervise two or more 'spotlights' at once. This ability is reduced with age – why we do not know. Perhaps the brain's information processing capacity decreases with time; this could cause one task to become mixed up with another if they are attempted simultaneously.

Awareness allows us to make decisions about the environment and discriminate between events. We need efficient and accurate attentional mechanisms to do this. In old people, these may be affected by prescribed medications. If an old person is less able to break down and excrete a drug then it will accumulate in the body. It may persist for so long that it is present in small amounts when the next dose is due. After several weeks of regular doses, drug levels may build up to the point where the brain is affected. If some degree of mental impairment was already present before the dosing began, then it may worsen to the point of mimicking dementia. This phenomenon is probably the single most common cause of readily reversible dementia in the developed world. It can have tragic consequences, including unnecessary admissions to hospitals and nursing homes, falls, and road traffic accidents. Set against these disasters, the effects of drugs on mental performance tests in studies of ageing may seem trivial. Unrecognised, however, they will serve to exaggerate the detrimental effects of ageing on mental abilities.

Chapter 4

Empty sieves and broken looms

How mental ability changes with age

Mental speed

Everyone who lives beyond the age of seventy will experience a slowing of mental speed. For some researchers, this slowing explains other deficits in memory and language. Mental speed is assessed by measuring reaction time. The simplest form of this test (simple reaction time) measures the interval between presentation of a stimulus and the response. The more complex 'choice' reaction time measures the time taken between the simultaneous presentation of several stimuli, discrimination between them and completion of the response. Choice reaction time slows with age even more than simple reaction time, pointing to involvement of brain structures in this phenomenon. Decreased nerve conduction speed, impaired sensory functions and slowed muscle movements are common features of both choice and simple reaction time. The likeliest source of age-related slowing is probably to be found in the mechanisms that transmit information from one nerve cell to another.

The averaged evoked response

Information processing by brain circuitry is difficult to study. The events we are interested in take place in groups of interconnected brain cells, last for a few seconds at most, and occur against a background of many other parcels of information being shared between brain structures. One ingenious method to study brain functions of this type is based on computer averaging of brain electrical responses – the averaged evoked response. Study of the averaged evoked response shows a clear decrease with age.

At rest, the surface of the brain shows considerable electrical activity. This is easily detected by placing two electrodes on the surface of the scalp and measuring voltage differences over time between the two electrodes. These will reveal a continuous, apparently random fluctuation in voltage over time. If one electrode is held in a position that is not over the brain, such as the ear lobe or the mastoid bone, and the other is attached to the top of the head, then a computer can help to decode the electrical activity taking place in the underlying brain. This is done by presenting a series of sensory signals. The signals can be auditory, visual or tactile, but in our example we shall use the hearing system. A pair of earphones are placed on the head and the subject is instructed to count all the high tones ('pips'); these comprise 15 per cent of all the tones heard that session – all the others are low tones ('bleeps'). The computer records the electrical activity for the second after each tone. This is amplified about one million times in the computer and subdivided into segments, each of 4 milliseconds' duration.

Typically, 200 short sounds will be heard by the subject, randomly ordered as high (30 pips) or low (170 bleeps). The computer will average the brain response for each 4 ms segment after all pips or bleeps. Figure 6 shows the typical averaged response to the pips ('oddballs'). Voltage waves above the zero line are termed 'P' for positive and those below the line are termed 'N' for negative. A positive wave picked up in most people at around 300 ms is termed the 'P300 wave'. These later-occurring components (around 300–400 ms) of the averaged response are believed to reflect decision-making by the brain as the subject tries to make sense and categorise the signal. ('Is it high or low? What number do I give it?') As the task is made more difficult in young adults these late components become slightly delayed. The brain structures that generate these late components are not exactly known, but are believed to include the hippocampus. There is now a huge body of data on ageing and the averaged evoked response. There is slowing of around 100 ms in the P300 wave in old people when compared with young adults. The effect is even greater in dementia. Here the rate of response slowing in progressive

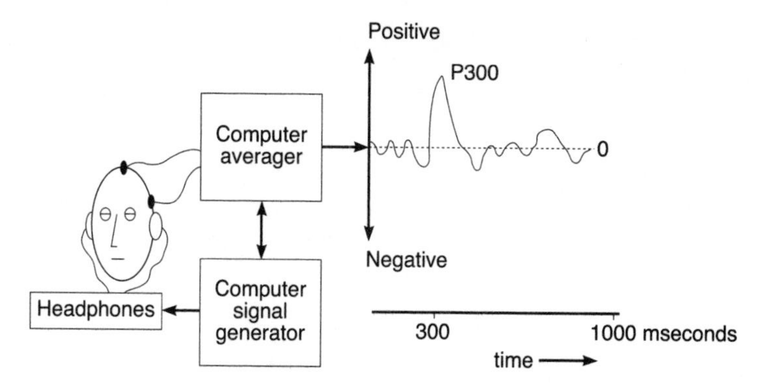

Figure 6 **The averaged evoked response.** The response can be 'evoked' by any type of stimulus – sight, hearing or touch. Here a computer generates a random sequence of noises that are mostly dull tones (80 per cent of the time) or clicks ('oddballs') 20 per cent of the time. The two electrodes on the scalp record electrical differences in voltage over each second (1000 ms) after each stimulus. After many hundreds of stimuli the computer will have stored and calculated the average response to a click and to a dull tone. The least frequent stimulus is examined most closely and shown here on the right. The large wave upwards at about 300 ms is called the P300 wave and is thought to represent processing of the stimulus by the brain. With ageing, this component occurs later, often around 400 ms by age eighty.

dementia can be linked to the severity of dementia. Although the hippocampus is confidently linked with the generation of the P300 wave, there is much less certainty about the site of slowing in dementia.

The executive controller

The chief executive of a large corporation is responsible for planning, initiation of new developments, coordination of responses to changed trading conditions, picking up cues from the market, and showing leadership. The brain must do these tasks or something very like them if it is to be successful (Figure 7). Studies of brain-damaged patients reveal that when the frontal lobes of the brain are damaged in adulthood, deficiencies quickly appear in planning, judgement and emotional control. When

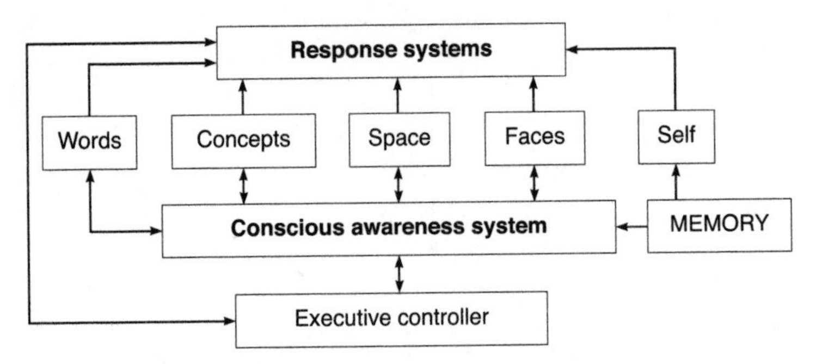

Figure 7 **The work of the executive controller**. The key box is labelled 'response systems' and this coordinates an individual's responses to internal and external needs. The executive controller interacts closely with a 'conscious awareness system' which both directs and receives information about the self and relies only partly on language to do this. Memory is much more complex than shown here.

tests of frontal lobe function are compared with tests of other brain areas, the test scores for the frontal lobes show the greatest deficits with age. These simple facts have encouraged psychologists to consider that the 'executive controller' in our theatre metaphor may become inefficient with age, resulting in impaired performance on memory tasks. In these terms, the 'executive controller' would have responsibility for the categorisation and manipulation of memory rather than its storage. This type of memory function is sometimes called 'prospective' memory. It provides strategies to remember to switch off appliances before going to bed, locking up on leaving the house, remembering birthdays, or taking tablets. More annoyingly, if impaired it may fail to provide a reminder that something has already been told. This may account for some old people repeating the same story to the point of tedium.

A successful company chief executive will demonstrate flexibility in changing market conditions: an operational mantra learnt at business school is useless when rigidly adhered to in the face of poor trading performance. Similarly, the brain's executive controller is expected to show flexibility, retaining the ability to generate new ideas and new solutions, and to remain creative across the life span. Tests of mental flexibility demand a great deal of the subject. Typically they begin with word generation tasks: for example, the person tested may be asked to name as many words as possible beginning with the letter 'S' within one minute. Almost all young adults can produce at least ten. Verbal fluency tested in this way is impaired after damage to the brain's frontal lobes and to a variable extent with ageing.

The task of the executive controller prompts a return to the first of the three questions posed in the last chapter: is brain ageing general or selective? If the executive controller becomes inefficient with age, does it follow that all mental functions are equally affected? It would be easier to answer this if the structure and location of the executive controller were precisely known. It is suspected to be based on neurons in the frontal lobes, but which ones is not clear. The speed and efficiency of the executive controller is crucial in mental life. Slowing and inaccuracy could cause many errors, possibly sufficient to explain psychological ageing.

Intelligence

What is intelligence? There is no clear answer. Most definitions seek to convey the idea that intelligence is the capacity to use powers of reasoning and judgement to make accurate appraisals of circumstances, solve problems and act with purpose in response to needs. These ideas about intelligence probably owe most to the requirements of psychologists, who want to detect and measure differences between people's ability to use their intelligence. This means making judgements about the reasoning and judgmental abilities of others. To do this, we must observe people performing tasks that we have selected because we believed them to be good measures of abilities that have also been chosen by us, because we thought they contained elements of our definition of intelligence. Critics of work on intelligence have had a field day with this type of argument. Some believe that intelligence cannot be measured because it does not exist. Others go further, and suggest that the invention of the tests caught the imagination of wealthy industrialists in the early twentieth century who needed a cheap and simple way to select youngsters who were most likely to benefit from a better education. These better-educated young men (they were always men) would in turn be trained to fill job vacancies as managers in industries with complex manufacturing processes. A cheap supply of people selected for their 'intelligence' would permit the rapid expansion of technology-based manufacturing industries, increasing profits and wealth without the need to improve education generally.

Intelligence tests were taken up with enthusiasm by the US Army in the First World War. Military authorities wanted to recruit the more able and reject those who lacked the mental ability for modern warfare. The army called the test the Army Alpha Examination, and gave it to about a million and a half potential recruits up to the age of sixty. It was quickly noticed that among the potential officers, older men did less well on the test than younger subjects. This gave rise to the idea, which persisted probably up to the 1950s, that all aspects of intelligence declined with age.

Intelligence tests developed gradually over the subsequent twenty years or so. Some researchers did not appreciate that Yerkes – the psychologist who had analysed the US Army studies and authored the report in the early 1920s that intelligence declined with age – was concerned that the result was more apparent than real. Yerkes fully understood that the sample of recruits compared a range of individuals born between 1866 and 1900, and was aware that factors such as improved diet and better education might account for differences between age groups in test performances.

Modern intelligence tests were further refined by David Wechsler during the Second World War. His tests distinguished between 'verbal' and 'performance' intelligence. Wechsler found that verbal scores were broadly similar in all age groups (from twenty to sixty) but performance scores were lower in the older adults. He interpreted his observations in much the same way as Yerkes had twenty years earlier, and concluded that intelligence declined with age.

By the 1950s, researchers were becoming more sophisticated in their use of different types of research design. Bill Owens thought the veteran survivors who had sat the Army Alpha test in 1917 and 1918 would be ideal subjects to test the then widely held view that intelligence declined with age. He followed up a group of men who had been nineteen years old and about to go to college when they sat the Alpha test. Their scores from 1917 were compared with retests taken thirty-one years later. The results were striking. The fifty-year-old subjects did not display the expected decline in intelligence scores. In fact, many of them showed overall improvement, while some elements (vocabulary, practical judgement, information and disarranged sentences) showed consistent gains with age. This study probably provides answers to the questions we posed earlier; we can begin to conjecture that ageing is not necessarily all bad for all mental functions, and that some abilities may in fact improve. We can suspect that ageing has selective effects on mental abilities, affecting some and sparing others.

When people challenge widely held views (dogma) in science they expect a robust rebuttal from authority figures in their particular scientific discipline, and this is exactly what happened to Owens. All sorts of reasons were put forward to explain his 'obviously erratic' results. Sampling error was seized upon as the likeliest explanation. Was it not likely, said some, that his sample was biased? Did not the sample include the initially more able, just about to get extra education intended to improve their performance and their later ability to learn from life? Others wondered if the survivors who had agreed to retesting were those who had been more successful in life. Eventually, the view prevailed that cross-sectional research designs have some value but are fundamentally flawed methods of examining changes in intelligence with age. The best approach is to re-examine the same individual in a longitudinal study, even though this will introduce the bias of test practice into the results. This change in thinking was accompanied by recognition that intelligence might not be a single measure of general mental ability but might have its own substructure. Different subunits of intelligence may each be affected differently by ageing – much as Owens had suspected.

From the 1950s onwards, a more complex picture developed of the structure of intelligence and the effects of ageing. The simple idea that a single measurement called 'intelligence' declined with age could not be sustained as more and more longitudinal studies were reported (Figure 8). This concept was replaced by the idea that there are two basic aspects of intelligence: 'crystallised' and 'fluid' intelligence. The term 'crystallised' describes the sort of knowledge, solutions and verbal abilities that accumulate with experience; the domain is represented by metaphorical 'crystals of knowledge' compacted by

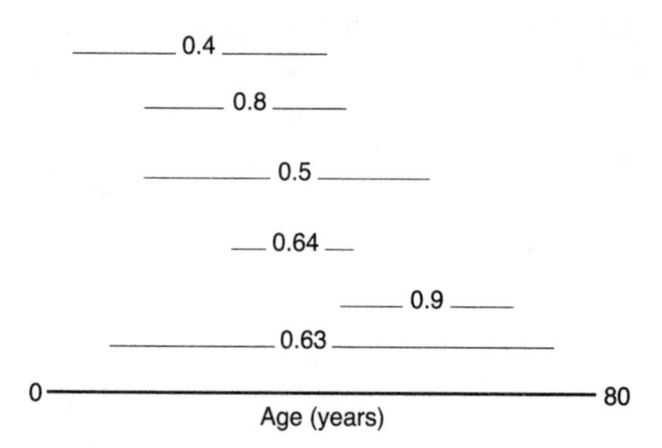

Figure 8 **Intelligence is stable across across the life span.** The figure summarises studies of intelligence tests used to measure stability of IQ test scores in the same people measured at two points in their life span. The longest interval was reported by Ian Deary and the author in one hundred people first tested aged about eleven and retested exactly sixty-six years later. The numbers (correlations) show the strength of the stability, the higher the number the more stable is intelligence. If intelligence was perfectly stable the score would be exactly 1.0.

time. In contrast, 'fluid' intelligence conveys the qualities of mental flexibility and willingness to adapt to new circumstances, to address novelty and to try to make an appropriate adjustment.

These ideas proved very useful in clinical practice. They are the first way of thinking about mental changes in old age that gives a simple framework for understanding the ageing mind and planning future studies. For many old-age psychiatrists, the distinction between 'crystallised' and 'fluid' intelligence is sufficient for most of their needs.

This simple division of intelligence into 'crystallised' and 'fluid' components has been supported by good evidence from cross-sectional studies showing that crystallised intelligence tends to be well maintained across the adult life span whereas fluid intelligence tends to decline. Although few longitudinal studies have tested the differences between crystallised and fluid intelligence with ageing, the idea has proved so durable and popular that many modern researchers remain content with the concept.

A minority continued to look for a new approach to the development of intelligence into old age. Their reasons were fairly straightforward. First is the fact that during the twentieth century the intelligence of successive birth cohorts was seen to rise gradually (the Flynn effect). The cause of this phenomenon is unknown. It is unlikely to be genetic and is probably attributable to environmental factors. For example, education is now more widely available and probably better structured to accommodate a wide range of mental abilities. The provision of health services is also better: less education is lost because of the chronic treatable diseases of childhood, and nutrition may have improved since the early years of the century. Second, the suspicion grew that the apparent decline in fluid ability during adult life may be attributable to the cross-sectional nature of the early studies. Aware that this type of research design had earlier prompted the conclusion that overall intelligence declined with age, and that longitudinal studies had provided strong evidence to the contrary, alternative explanations were sought. Perhaps the Flynn effect was more marked for fluid than for crystallised abilities? If so, cross-sectional studies would show a decline in fluid and a retention of crystallised intelligence, but longitudinal studies would not. Careful follow-up studies showed that fluid intelligence did in fact decline much as expected, but that crystallised intelligence scores were maintained.

Others wondered if the decline in fluid ability was the result of disuse. Much as muscles shrink if a once-active athlete ceases to train, could brain power be similarly affected? 'Use it or lose it!' became an occasional but lively contribution to debates on the subject in the early 1980s. This idea is of more than just academic interest. Serious attempts were made to remedy deficiencies in fluid ability by intensive training of old people, to the extent that some did indeed improve their abilities. Other researchers were interested because they wished to avoid some of the costs in time and money incurred in longitudinal studies. They believed that more rapid progress could be made if the decline in fluid ability could be reversed by training. The argument went as follows: if this decline could be reversed, then it was not an inevitable consequence of ageing. Most doctors agreed with this approach, without concerning themselves with the intellectual argument. Some studies have shown that familiarising an old person with a test of fluid ability will improve performance on that test. Critics say this is not a relevant argument and attribute any improvement to a 'practice effect'. However, supporters reason somewhat differently, suggesting that anything that can be shown to improve fluid intelligence is of potential therapeutic importance. Furthermore, they argue that the fact that these practice effects can be detected three years or more later is certainly of relevance to the individuals concerned.

Studies of this type are also criticised from a standpoint similar to that taken by the early critics of intelligence tests used to select children for different types of secondary education, as was done for many years in the UK – the dreaded 'eleven-plus'. At that time, the worry was that the tests were unfair to children from the type of home in which there was little time for reasoned discussion. When verbal explanations are unforthcoming, where there are few opportunities to spend time with paper and pencil or solve puzzles without distraction, then children may be disadvantaged in a formal intelligence test setting. The same children may, however, develop appropriate social skills to solve other types of problem and adjust comfortably to their circumstances. This line of argument suggests that intelligence tests are not honest measures of the intelligence of old people if the tests fail to examine the aspects of mental life that old people use every day to maintain independence, keep in touch with friends, and so on. Critics conclude that intelligence tests are useful for young people but much less so for the old. Some go further, and emphasise that when a test designed for young people is applied to the old, any deviation in performance from that expected of the young is interpreted as a deficiency of old people.

The usual educational attainments of older birth cohorts are much less than those of younger cohorts, who may be trained in more abstract ways of thinking and problem-solving. This may be part of the reason for older people doing less well on intelligence tests. An alternative approach would be to find within a single birth cohort just what distinguishes the mental performance of one old person from another at tasks old people themselves see as important in day-to-day life.

Most old people agree that it is their ability to use their acquired knowledge (crystallised intelligence) to solve everyday problems that they rely on most in this context. In late life these abilities are closely linked to the emotional processes that scrutinise new experiences before selecting some to be placed in context of their overview of life so far. This integration of feelings and knowing which takes place slowly in old age is recognised and explored throughout our history. Previously referred to as 'wisdom' and more recently as 'ego integrity', the integrative processes of the wisdom of old age are brought to bear on questions concerning the purpose or consequences of our actions. Wisdom is not intelligence, which brings the powers of analytical thinking to problem-solving. For many the difference between the two is that intelligence is used to answer the question concerning how something should be done, while wisdom asks whether it should be done at all.

One of the most controversial aspects of the study of intelligence concerns theories about its structure. As a general rule, the more items a mental test contains, the better it will be as a measure of a mental ability. If at the outset we agree that a range of mental abilities needs to be tested in order to estimate a person's general level of intelligence, then an intelligence test must contain many items for each of these abilities. Charles Spearman was a leading British psychologist who became interested in intelligence in the early twentieth century. He wondered if a person's performance at one type of test (say, mathematics) could be vastly different from another (say, verbal reasoning). To tackle this type of question, Spearman used a statistical method (factor analysis) to observe relationships between large numbers of items in intelligence tests. This allowed him to look at the question of relationships between groups of items and then how much performance on one group relied on another. Spearman concluded that there is one large general factor ('g') which could account for differences between individuals. Later workers developed other ideas. Some suggested that a small number of important abilities could best explain relationships between items; others suggested that intelligence comprised an enormous personal archive of individual solutions to problems old and new. A consensus does not as yet exist. Spearman's position remains attractive; it offers the possibility that *g* will be linked to biological components (such as measures of mental speed). In turn, these biological factors may prove modifiable in late life and so support retention of mental abilities across the life span.

Language, memory and ageing

There are many positive gains from growing old. However, this is rarely the view of young people, who see few advantages in old age. What appeals to the young may be beyond the reach of most old people. Going out and enjoying yourself with new friends requires the means, the motivation and also the opportunity to do so – which seldom come along at the same time for most old people.

The idea that the brain continues to develop as it ages is quite new. Until recently, scientists working on how the brain makes and maintains cell-to-cell connections thought it was useless to include older animals in their studies because they assumed that the capacity to make these changes is lost as we age. The brain's first response to the slowing of mental processes caused by ageing is to compensate by using all the mechanisms of synaptic plasticity at its disposal.

Plasticity – a concept that was introduced in Chapter 2 – can be found at many levels of brain function. It is evident not only in changes in dendritic branching but also in biochemical alterations in the functions of large regulatory biomolecules. In later chapters you will read about the brain's 'self-organising principle'. There is a well-founded suspicion by researchers that in response to ageing some brain functions are more capable than others of this sort of reorganisation: this would explain why some mental abilities are better retained than others with ageing. The exciting part of research in this area is the prospect of discovering the sources of these differences. If they prove to have a biological explanation (and many suspect they will), then we may be close to discovering how to retain our mental abilities as we age. Perhaps we will be able to learn lessons from the preservation of mental abilities that are well retained and apply them to those that are lost.

Case 1

Jim A. was a successful businessman who had made several fortunes before retiring, becoming bored, starting a new business, retiring again and starting yet another business before finally selling all of his interests at the age of seventy-four, at the insistence of his wife, who wanted him to travel with her. He was a larger-than-life character with a considerable capacity for personal warmth and for holding the attention of those around him. He was a little bored by his wife's travel plans and spoke at our first meeting about symptoms that sounded at first rather like a depressive illness. He said he missed the thrill of making deals, outsmarting younger and quicker minds, and was particularly annoyed that he had been obliged to give up his private pilot's licence. He had no other symptoms suggestive of a depressive illness, and as the interview progressed it became clear that his main concern was his failing memory. He had first noticed this some years earlier when his personal assistant had retired and he realised how much he had relied upon her to organise his day and guide him through social functions. Now he felt he could not easily put a name to a face, often forgot important appointments, and found many personal financial matters were much more efficiently dealt with by his wife. Testing his memory revealed no difficulty in paying attention or in the registration of new information, but considerable difficulty in recalling new verbal information, especially after a brief interval during which he undertook another task. His easy charm covered up these deficits with pleasantries and little excuses. He could not conceal by the end of the interview how concerned he had become about his failing memory. He wondered if he was now certain to be demented. Brain imaging revealed shrinkage of the cortex and some expansion of the ventricles reported as unremarkable for his age.

There are different types of memory: long-term memories are laid down over many months and years and remain accessible throughout much of life, even in the latter stages of dementia; short-term memory is the current account of daily living. We use the term 'working memory' to show that there is some form of temporary storage in the acquisition of new information. When we try to understand something new or solve a puzzle we need to handle new information acquired perhaps over the previous few seconds or minutes. The new information, captured selectively by our sensory systems through sight, sound, touch and smell, is placed in the temporary short-term working memory store, where it is available for rehearsal purposes, encoding immediate decision-making, and for the modification of that information using various retrieval strategies. The short-term store can provide data for an immediate reaction in response to very recent sensory input. Other data can be transferred into the long-term store.

The components of working memory are fairly straightforward. First, there is an attentional control system – the central executive – served by two slave systems, responsible for the intelligent organisation and modification of information. One system is visual (the visuospatial 'sketch pad') and other is verbal (the phonological loop). The central executive has limited capacity and is a complex decision-making structure made up of many subcomponents. It can select appropriate strategies and plan what to do with information.

There is much overlap between the mental processes involved in memory and language, although this is more obvious in naturalistic settings than in the psychological laboratory. This overlap has been little studied owing to the tendency of researchers to examine memory and language in quite separate experiments. Our ignorance of this area does not stop psychologists from proposing that some of the memory problems encountered by old people might be traceable to a fundamental loss of the ability to understand language. Patients with an acquired disturbance of language will do less well on tests of memory even when the test does not rely on verbal material. The tester usually assumes that the disturbance of language is not the primary cause of the memory problems. Some take the opposing view, that memory problems in old age impair the understanding and production of language.

The next section looks at the usual problems of memory and language as we age. The two are intermingled in daily life but can be separated in the psychology laboratory. This separation may not fairly represent what happens in more natural situations.

Language and successful ageing

When we looked at intelligence changes with ageing we saw that the store of verbal knowledge increased with age. We know more words as we age, and very few old people complain of under-

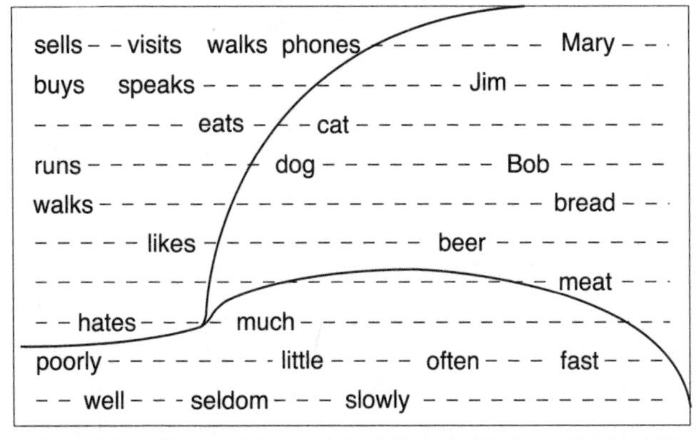

Figure 9 **The brain acquires language following built-in rules** and these break down with ageing. Every child is born with a capacity to acquire and use language. Brain cells that do this work are identified in development as well-ordered projections from one layer of the cortex to another (key areas are shown in Figure 3). These cells are sensitive to specific features of each spoken word. These cells are not arranged as in a map in the conventional sense – it is not a case of one cell per word – but as abstract maps of word features. These maps can be generated by computers and Figure 9 shows how the brain can organise itself to produce such maps. With ageing, it is the rules which hold the maps together that are weakened and not simply the loss of brain cells.

standing fewer words than when they were young. Some people experience a problem with words being on the 'tip of their tongue' – that is, they have a problem accessing words. If we listen carefully to old people with this type of problem we will hear other mistakes in their use of language. Some substitute the word they are looking for with a circumlocution which uses several words when one will do (for example, 'a thing for tidying my hair' instead of 'hairbrush'). Occasionally, just one word is used, describing the function of the object rather than the object itself (for example, 'cutter' instead of 'knife'). Mistakes of particular kinds are also frequent, so someone might ask at table for 'loaf' and not 'bread', or ask for quite the wrong thing, calling for sauce when soup is served. Ask these old people if they know what they mean, and you will not be surprised to be told that they know *precisely* what they mean – they just can't think of the right word.

Long, complicated sentences pose a different sort of problem for old people. True, many old people don't care for long-windedness and their own speech becomes less complex with age, but this is different. Old people, especially those over eighty, have difficulty understanding grammatically complex sentences. It is not known if these difficulties are caused by disturbances in the brain circuits that support language, or whether it is something non-specific. Slowing of information processing speed, impairment of attention and less efficient memory retrieval may be sufficient alone or together to explain language problems in old age, including the 'tip of the tongue' phenomenon.

The extent of language problems in old age must not be overstated. More than any other aspect of mental ability, language skills remain intact into later life. Stroke and dementia are much more important and extensive causes of language impairment than normal ageing.

Memory and successful ageing

Some slight decreases in long-term memory efficiency are widely reported in all studies of old people, and probably reflect exactly what is happening in the daily lives of their subjects. The laying down and retrieval of memories from long-term storage are implicated in many of these studies. Old people appear less capable of efficiently managing all aspects of memory processing. There are some important differences between types of memory. Implicit (also known as 'non-declarative') memory occurs when the subject is unaware that learning is taking place. Explicit ('declarative') learning happens effortfully: the subject is deliberately trying to learn when explicit learning takes place. Explicit learning always involves the hippocampus, whereas diverse brain circuits can acquire implicit learning. Age-related decline is most marked in explicit learning and memory. In some classic conditioning experiments older adults require longer to attain the same degree of proficiency as younger adults.

All old people show some decline in their ability to retrieve new memories. For most this is just a mild nuisance, but for others it becomes a source of considerable worry. Each year a small proportion (probably fewer than 2.5 per cent of those aged sixty-five to eighty) of old people with mild memory problems will develop much more severe memory difficulties, leading eventually to dementia. The majority of old people will not become demented, and mild difficulties are not recognised as a public health problem in the same way as, say, Alzheimer's disease. Also in contrast to the dementias, little is known about the brain processes that cause mild types of memory impairment. This ignorance of the biological basis of mild memory impairment is the major obstacle preventing proper drug development in this area.

Discussion of the role of the central executive and the modular theory of brain organisation prompted speculation that the deterioration in some brain areas (especially the frontal lobes) might explain specific psychological changes. Subsequent examination of complex aspects of higher-level problem-solving ability, language and memory emphasises the widespread patterns of neuronal connections required to support these mental activities. The three questions posed in the last chapter can now be re-addressed.

Is brain ageing general or selective?

There are a small number of general changes in mental ability. These include slowing of mental speed, slowing of information processing, and reduced ability to divide attention. Some aspects of intelligence are probably reduced in a fairly non-specific fashion. These are linked to mental dexterity or fluidity and mental flexibility. More research is needed to determine if these general changes can account for the selective changes in memory and language.

Do brain ageing changes occur at random?

There is no evidence for this from the results of studies of changes in mental ability. Patterns of change in mental ability are so consistent between individuals that random changes could not explain them.

Does ageing drain the brain's overall capacity?

This remains uncertain. Some of the explanations for age-related changes (in divided attention, for example) could be explained using this model.

Chapter 5

Keeping your nerve

Ageing, stress and the brain

Some years ago the British government set targets to improve public health, published as *The Health of the Nation.* These proposals became notorious among medical researchers for many reasons. Not least was the exclusion of priorities to improve the health of old people. A moment's reflection reveals the weakness of that government position. Old people are more likely to suffer disease. We all expect to be relatively disadvantaged by age in terms of mobility, sensory awareness, loneliness and (for many of us) poverty. Cynics explained the government's neglect by emphasising how civil servants had advised against measures that might increase the number of people needing health care, because this might add to the burden of care by the state. This is something of a defeatist position. Of course old people suffer more ill-health; but the correct response is not to ignore the problem. Rather, the question should be why ill-health is thought of as an inevitable consequence of growing old.

Old people are less efficient in their responses to rapid environmental change. Minor upheavals for a young person become major stressful life events for the old. These changes in adaptive capacity characterise the ageing individual: indeed, some physiologists and psychologists think the loss of this adaptive capacity is the best definition of ageing. Internal regulatory systems become inefficient, and the initial response to stress by an old person is greater than that of a young one. When the stress stops, the older person will take longer to return to baseline levels of activity.

Dying of a broken heart

In the early 1960s, Dewi Rees and Sylvia Lutkins were family doctors working in central Wales. Like many before them, they noticed that it was not uncommon for people to die within six months or so after the death of a spouse or a close relative. Unlike others, they decided to test if these deaths exceeded chance. They compared the mortality of 903 close relatives of 371 people who died, with 878 close relatives of 371 control subjects who did not die. The control subjects were matched for age, sex and marital status with those who died. During the first year after death, 7 per cent of bereaved close relatives died, compared with less than 1 per cent in the control group. The difference could be attributed chiefly to surviving spouses, whose mortality rate was about ten times that of the control group. These results were widely discussed at the time, but remain unexplained.

Working in a London teaching hospital, Murray Parkes took notice of the work by Rees and Lutkins. He showed that widowers were about one and a half times more likely to die in the first six months after the death of their wives than could be expected by chance. Of course, a husband and wife might share similar unhealthy behaviours (e.g. smoking) or poor housing (e.g. overcrowding) and so might share an increased likelihood of premature death. Another possibility is that the mental processes following bereavement may somehow increase the chance of dying. When these reports were first published no explanation was agreed, but considerable speculation persisted for many years in the scientific and popular press. In some quarters the finding was linked in the public mind with popular ideas about the 'will to live' and 'mind over matter'.

At about this time the concept of psychoimmunology had begun to gain acceptance as a valid area of scientific study. The term was used to describe the interaction between stress, personality, immune system dysfunction and the development of disease. The diseases of particular interest were either mediated through the immune system (such as allergies) or resisted by it (such as cancer or infection). A role for the brain in regulation of the immune system was supported by the work of Stein, who completed an elegant series of studies in the 1950s. He showed that some, but not all, surgical lesions in the rat brain could impair immune function. These decreases did not impress contemporary immunologists, who found it difficult to relate slight but statistically significant reductions in immune function to the much greater degree of immune suppression typically associated with the development of disease. This effect-size problem has bedevilled studies of the immune system and its possible interactions with mental processes ever since.

Is the immune system involved in brain ageing?

The immune system is composed of two essential components. The first, 'cellular' immunity, is largely the function of white blood cells (lymphocytes) that can act directly upon an invading antigen. 'Humoral' immunity is provided by white blood cells that can produce antibodies. The immune response to an extracellular antigen such as a bacterium is mounted by the complex interaction of the two systems, each comprising three elements ('the two trinities of immunology').

The major immune cell types are B cells (derived from Bone) which secrete antibody to neutralise the invader; T cells (derived from Thymus) which function by direct contact to kill cells attacked by an invader; and a group of cells known collectively as 'accessory cells', which latch on to foreign invaders and present them to the T cells. During ageing, each of these cell types shows a consistent pattern of increase in some of their functions and decrease in other functions. Ageing of the immune system does not reflect a simple decline in the efficiency of immune surveillance, but rather the disorganisation of a highly complex and effective protection system during late life. Some types of infection, for example by viruses, produce a typical response from the adult immune system. In old age, this response is slower and less effective.

Damaged cells release chemicals which provide signals to the immune system. The balance of these 'messenger' molecules alters with ageing, so that molecules that stimulate an inflammatory reaction tend to increase, and those that enhance immune function tend to decrease. For some researchers, these changes in the immune system with ageing are attributable to collapse of the thymus gland. This gland is situated in the upper chest and is responsible for the maturation and development of T cells. Reduced function of the thymus gland causes a slowly progressive decrease in the proportion of native T cells (not yet differentiated in response to an antigen) to memory T cells (cells modified to respond rapidly to a repeated challenge). Within the memory T cell population, there are many age-associated defects at a molecular level in the sending of messages within the cell and signals to stimulate the growth and division of T cells. In old people there are T cell deficits in the response to virus infections and reduced ability to identify foreign material. In later parts of this book we will look again at the inflammatory response in the brains of old people and ask if anti-inflammatory drugs can help prevent dementia.

Of course, these ageing effects in T cells could simply be part of non-specific changes in other cellular systems of the body. This idea is part of the theory that all cells in the body will eventually become affected by ageing changes ('cellular senescence').

The immune system has been investigated as a possible mechanism to account for the increased death rate following bereavement. Schleifer and his team measured the ability of white blood cells to divide in response to a stimulant of cell proliferation. Fifteen married men were studied during and after the death of their wives from breast cancer: T and B cell levels in these men remained unchanged throughout this period, but the cellular response to stimulants of cell division was reduced. This and many similar studies were reported during the 1980s but their relevance has not been established. The jury remains undecided.

Is growing old stressful?

A pioneer of stress research, Hans Selye, has implicated the biological responses to stress as an important component of body ageing. For many old people, old age is indeed stressful, for reasons that are not difficult to see. Reduced resources are available for the old person to muster in response to a new demand, a novel task or an urgent request. This is because the total capacity to complete the task is reduced with ageing and the effort required may bring the old person close to a level at which mental or physical systems can no longer cope. This phenomenon is well known in studies of ageing organs such as the lungs and heart, and is known as 'decompensation'.

The main physical changes in body systems associated with ageing are best understood in terms of 'diminished functional reserve'. Optimum or peak organ performance is established in early adult life, and thereafter there is a gradual decline in capacity. In the face of a stressor, the reserve capacity must be adequate to cope with the increased demand, otherwise symptoms of decompensation will arise.

Psychologists also accommodate the concept of 'functional reserve' in their development of mental tests for old people. Too much pressure to complete a mental test and what seems a simple task may trigger a 'catastrophic' reaction, in which the old person is overwhelmed by a sense of failure, hopelessness and a need to escape from the test setting – a response that would have been unthinkable for the same person in early life or middle age. Ageing, however, redefines and reduces the stressors needed to produce a stress response. Where once an individual might have coped efficiently with demanding tasks, with age such tasks become effortful and eventually are carefully avoided.

The stress response

A stressor is any change in the outside world that perturbs the constancy of the internal physical or mental world. How the body copes with these external changes is termed the 'stress response'. The response comprises a repertoire of brain and hormonal efforts to adapt to the stressor. In psychological work on stress, it is also necessary to include stressors that arise not from the outside world but from within our mental life. Some can be as simple as anticipation of the imminent arrival of a stressor: a good example is the sense of fear one feels before sitting a test. It reflects awareness that if one remains in a particular setting, then a stressor will arrive. For the old person, this type of stress can be accompanied by the sense that one is too old and weak to run away or defend oneself. Other psychological stressors are more complex, and represent attempts to cope with internal drives or feelings with a sense of pervasive violation of true and lasting values.

To understand the stress response, the part played by the brain and how the ageing brain leads to impairment of the stress response, it is helpful to look at how physiologists reached their present understanding. The nineteenth-century French physiologist Claude Bernard was the first to consider the stress response. He saw that animals had evolved from lower life forms that probably developed from the first single organisms to be independent of the external environment. Internal cellular chemical processes, he noted, require precise chemical conditions if they are to perform at optimum efficiency. Bernard proposed there are internal cellular regulatory processes to maintain these chemical conditions. He coined the term 'the constancy of the internal milieu' and thought that the purpose of this system was to buffer the internal cellular environment in the face of abrupt, rapid and possibly hazardous environmental change.

Claude Bernard's work was later built upon by the American physiologist Walter Cannon. He wrote widely and persuasively for the general public about modern physiology, in works such as *The Wisdom of the Body.* Cannon introduced the term 'homeostasis' to describe body processes that could maintain a constant internal environment. Later, Hans Selye recognised that the body's response to stressors was directed towards adaptation and introduced the term 'general adaptive response' to cover all components of the stress response. He also saw that the hormones produced by the adrenal gland were an important component of the stress response. (The adrenal glands are a pair of small glands about the size of a walnut, each shaped like a three-cornered hat and sitting on top of each kidney.) Selye noted similarities between the effects of exposures to chronic stressors and to patterns of age-related degeneration. With this in mind he and his colleague Tuchweber developed in 1976 a multiple stress hypothesis of ageing.

The role of hormones

There is a particular disease of the adrenal glands where excess production of cortisol causes a wide range of changes in almost all body systems. (Cortisol is one of the steroid hormones produced by the adrenal cortex to regulate the metabolism of sugar, mineral balance and reproductive functions.) Excess cortisol production is linked to muscle wasting, thinning of the bones, hardening of the arteries, diabetes, dysfunction of the reproductive system, decline in immune system efficiency and an increased incidence of cancer. Selye noted that many of these harmful changes also occur with ageing. The work of Selye and colleagues was much influenced by studies published about twenty years earlier showing that ageing in some fish and mammals was linked to excess cortisol production. The life of the Pacific salmon is normally terminated after first spawning by the effects of excess cortisol activity. Similar effects could be reproduced experimentally by injection of cortisol.

These first researchers were intrigued by the possible link between cortisol secretion and ageing, and wondered if cortisol production was linked to stress or disease. It prompted ideas that cortisol production might affect the body in some sort of cumulative way. In lower mammals this is certainly the case. In almost all conditions studied (rest, mild stress or substantial stress), exaggerated increases of cortisol secretion are reported with ageing. Not only does ageing increase overall cortisol secretion, it also delays the return to baseline cortisol secretion after exposure to a stressor. In old men and women the adrenal cortical system is much more difficult to study than in lower animals, largely because the system is so responsive to stressors.

Secretion by the adrenal glands is controlled by the brain. A substance known as corticotrophin releasing hormone (CRH) is released by brain cells and stimulates secretion of adrenocorticotrophic hormone (ACTH); this in turn stimulates the adrenal gland to produce cortisol. One of the effects of cortisol is to reduce the secretion of CRH by the neurons, in effect suppressing its own production; this elegant form of hormonal control is called 'negative feedback'. Ageing disrupts this negative feedback; there is also long-term loss of negative feedback inhibition of the cells controlling the adrenal cortex. This loss is probably involved in both the overall increased background secretion of cortisol with ageing and the delayed return to baseline in cortisol secretion after exposure to a stressor. The important neurons involved in this negative feedback loop are located in the hippocampus. It has been known for years that surgical lesions of the hippocampus increase stressor-induced cortisol release and that the hippocampus contains high concentrations of receptor molecules which can recognise the various types of cortisol produced by the adrenal gland. In ageing animals, the hippocampus becomes gradually less and less able to recognise circulating cortisol. This is probably linked to an absolute decrease in the number of receptor molecules. Behavioural studies with aged animals indicate that this can be linked to the number of brain cells lost from the hippocampus and the ability of the animal to learn new information.

Two types of cortisol are released by the adrenal cortex: one is largely involved in the regulation of sugar metabolism (glucocorticoid) and the other is involved in salt balance (mineralocorticoid). At present it is uncertain which of the two types is involved in disruption of the ability of the hippocampus to regulate cortisol production. What is important is that this dysregulation of cortisol production with ageing is certainly implicated in the impairment of brain function with age. In general terms, as the brain ages it becomes less and less able to regulate cortisol production, and the brain cells involved in the hippocampus become more vulnerable to damage. Animal studies show that both types of cortisol can influence many aspects of brain function. These include the role of chemical transmitters, learning, and biological rhythms. It remains unclear, however, whether corticosteroids are involved directly in toxic damage to these brain cells, or whether they are more readily understood as 'permissive' agents.

Changes that occur in the ageing brain, such as reduced synaptic plasticity, reduced axon sprouting and reduced dendritic remodelling, can also be induced in young animals by treatment with glucocorticoids. Chronic stress has a similar effect on overall aspects of ageing in the hippocampus. Many of these actions of glucocorticoids on the brain can be reversed by treatment with specific glucocorticoid antagonists. The harmful effects of stress are probably mediated through cortisol production. Certainly, essential cell structures such as the cytoskeleton can be made more vulnerable by cortisol to the sort of degeneration induced by a stroke. Acute stress probably induces a cascade of neurodegeneration that involves impairment of glucose transport into brain cells, destabilisation of calcium balance and greater production of 'reactive oxygen species'. The enhancement of oxidative processes in the older brain by glucocorticoids has the potential to accelerate the normal ageing process and is probably implicated in neurodegenerative disorders (like stroke) that involve free radical-mediated damage.

Cortisol production is involved in abnormal changes in the brain with ageing as well as the exaggeration of normal ageing. Although much of the evidence comes from animal studies, increased exposure to glucocorticoids in old age is thought to lead to brain cell shrinkage and ultimately to the death of brain cells involved in memory. One of the consequences of excessive exposure to glucocorticoids is that the branching of dendritic trees becomes impoverished. To compensate, the remaining undamaged nerve cells begin to sprout in order to restore dendritic spine density and length, and to renew or replace now empty synpatic contacts. This process is called 'compensatory synaptogenesis'. With ageing, compensatory synaptogenesis occurs much more slowly, probably because glucocorticoids can suppress neuronal sprouting in response to local damage. Animal experiments suggest that in addition to involvement in brain cell damage, glucocorticoids may actually reduce the compensatory response of unaffected neurons, probably by suppressing the genes involved in synaptic repair.

Unfortunately, studies in humans are less convincing. However, it is clear that differences between individuals are much greater within groups of old people than within groups of young people. Increased baseline cortisol production occurs in some, but not all, old people.

Psychological effects of cortisol in ageing

Physicians working with patients who suffer from overproduction of cortisol or who are given large doses of cortisol to treat medical conditions recognise a temporary and reversible psychotic illness ('steroid psychosis'). This disorder includes impaired memory, impaired attention and difficulties with reasoning. The extent of these difficulties relates quite closely to the blood concentration of cortisol, and in several reports older people were more severely affected than younger ones. Studies on human volunteers indicate that the general psychological problem caused by cortisol starts with an inability to discriminate between recently presented bits of information and irrelevant information presented as a distractor. This suggests that high levels of glucocorticoid in blood probably first affect the process of selective attention, then reduce the ability to acquire relevant new information; as the blood concentration of cortisol increases, memory is also impaired.

Observations of this type lead to an obvious question. Are the memory impairments thought of as part of normal ageing caused by increased stress responses in old people? Selye had previously suggested that ageing could be accelerated by exposure to stressors. Sonia Lupien and her colleagues saw that in response to stressful or non-stressful situations, test performance could be linked to memory impairment, supporting the idea that some part at least of normal cognitive ageing could be linked to aspects of the test situation.

Stress and ageing are linked to memory impairment

Stressful experiences are very individual. For one person stressful events are awful, never-to-be-repeated experiences; another person will emerge 'steeled' by stress, hardened into a stronger person. The nature of the experience depends on personal evaluation of the stressor, the choice between available responses, as well as some self-assessment of the chances of success. Key brain structures such as the hippocampus are involved in setting up a stress response, and the same structures are also crucial in memory. They probably process a great deal of emotional information as well. These brain structures are the probable source of the differences between individuals in their response to stress.

In humans the stress response does not typically involve the animal 'fight or flight' response. We are much more sociable than that. Instead, most of our stress responses are triggered by interpersonal events such as a difficult situation with a work colleague. Humans are very vulnerable to the effects of minor stressors that crop up repeatedly. For us, there is no option to fight over a mate, or to kill for food or territory. Our 'civilised' alternatives tend to maintain long-term hormonal and nervous arousal – valuable for an animal about to fight or flee, but in humans leading to cardiovascular hyperactivity and gut motility with the very real hazard of the development of chronic disease.

The ways in which ageing, stressful experiences and emotional life are linked together in the mature individual suggest an emerging pattern of personal stress management that involves influences as diverse as genetic makeup, dietary habits, earlier success in handling stress and the ability to learn from experience. This pattern may succeed until ageing depletes the available reserves, be they emotional or physical. Previous excesses or privations may have led to preclinical disease (such as hardening of blood vessels, linked to poverty, chronic stress and smoking). In old age the stress response may thus directly cause something as catastrophic as a heart attack or a stroke.

Chapter 6

Tired of living and scared of dying

Coping with ageing

In Chapter 5 we looked at the effects of stress on the ageing brain, and the impairment of immunity in older people. These issues are also relevant to understanding the psychological changes that occur in ageing – how the map of the mind is altered by the stresses of growing old and the increased vulnerability to disease. These changes begin well before old age is evident, and are often dealt with using coping mechanisms that were acquired much earlier in life.

Disengagement

Many men and women work increasingly hard during late middle-age – often in the face of hostility from younger colleagues who openly consign older workers to the employment scrap-heap ('useless after forty and nine o'clock at night!') At work, at home and in the wider community, people in late middle-age are doing their best to make good decisions, to broker peace when asked to resolve conflicts, and – despite overt criticism from younger people – to remain positive in attitude and purpose. Eventually it all becomes too much. Parents die, children leave home, job horizons narrow and there are health worries. Soon the anticipation of one's own mortality seems to pervade more and more of the working day. 'If I had the chance, of course I'd go – just like that ... I'd never wear a suit again.' Although there are always exceptions, it is true for many as they approach the end of their working lives that work ceases to be the focus of those lives. A process of disengagement begins which succeeds in reducing some of the stressors, lifting some of the burden of unpleasant feelings about work.

In people who have a lifelong strong commitment to work (not necessarily to a single employer or single type of occupation), important aspects of self-image fuse with their work status. This occurs in both positive and negative senses. Loss – or threatened loss – of that job produces an emotional reaction proportionate to the intensity of the commitment: strong commitments – when threatened or severed – produce strong emotional responses. These strong emotions change bodily feelings and alert the ageing individual once more to the problems of mortality and personal frailty. This is the basis of our presumed link between stressful life events, emotional health and bodily well-being. The link is so important that it has been extensively studied in medicine, psychology and sociology. It is soundly based on the all-too-common scenario of the middle-aged man who suffers his first-ever episode of severe illness. 'It was the stress he was under, you know ... terrible the way that they treated him ... I think they just wanted rid of him.'

Stressful life events

The study of stressful life events begins with the work of the psychobiologists of the early twentieth century. These pioneering psychiatrists and psychologists were interested in individual differences in response to adversity at all stages of the life span. Why did some people remain resilient in the face of extreme privation or stress, whereas others just couldn't cope? Researchers have attempted to grade stressful life events, giving them a mathematical weighting according to their capacity to induce a stress response. Experiences such as divorce, moving house, the death of a spouse and job loss were ranked particularly highly.

This type of analysis may seem somewhat irrelevant to the study of ageing, until we realise that stressful life events become more frequent in our lives as we grow older. A negative experience such as losing one's job may be concealed by a negotiated early retirement; it is more difficult to avoid the emotional consequences of children leaving home, and of age-related illness in a spouse or life-threatening symptoms in oneself – all experiences that become more frequent the older we grow. Likewise, our emotional commitments become stronger within loving relationships as these mature. The sudden loss (or threat of loss) of a loved spouse in late middle age or shortly after retirement can leave the survivor bitter and cheated of that part of their lives together for which they had both laid such careful and pleasurable plans.

Free from worries about work, children and presuming lasting good health, some can approach the prospect of old age with pleasurable anticipation. However, this is not generally the case. The risk of encountering a stressful life event is not evenly spread amongst us. Poor conditions in which wage-earners often work long hours, sometimes at heavy manual work, where living spaces are overcrowded or hazardous, without easy access to transport, are linked with ill-health and premature death in almost every society. Economically disadvantaged people tend to die younger. The more affluent we are, the longer we live and – typical of the unfairness of life – the fewer and later are the stressful life events encountered.

The social background to the stresses of old age

Societies differ in the type of event found stressful by old people. These differences relate to the place of the old person in that society, the level of social integration provided for old people and supported by values within the society, the point in the economic development cycle at which the person enters old age, and the opportunities provided by society to assist old people continue personal development in the last decades of life.

Advancing age is accompanied by a general decline in muscular strength, which deteriorates steadily in post-menopausal women and in men after the age of about sixty. Exercise and training will build up muscle tissue throughout much of life, but muscle speed and stamina are not so readily restored. Physical frailty imposes a much greater burden on someone whose work and role in society has relied on physical strength: loss of physical power will impose more hardship and unwanted dependency on a manual worker than on a sedentary worker like an office clerk. This is recognised in the occupational pension schemes for UK workers in the emergency services, but for other largely manual workers there is often neither assistance nor advice available when planning for old age. Likewise, the chance of suffering a disease in old age is much greater in poorer than in more affluent communities.

The stress of bereavement

Earlier we saw how bereaved spouses were at greater risk of premature death in the first months after their loss. Almost every study of old people establishes bereavement and loss as the major stresses of late life. Some widows and widowers feel that they have never recovered from the loss of their loved one. Mortality from physical disease – often related to the vascular system – and from suicide is higher than expected for two years following the loss of a spouse. Nevertheless, interviews with old people point to psychological resources that can be brought to bear on such losses. Because the loss is often age-related, and in an important sense expected, acceptance of the loss can sometimes be easier than in younger people, although the process is no less painful. Stoicism is said to be greater in older people, but this usually means the emotional costs are borne internally, without an overt display of distress. An old person who regards a partner's death as inevitable may still follow a path of emotional withdrawal and disengagement from society as a consequence. The sense of such a response is self-evident. Why make and maintain warm emotional bonds, only for them to be painfully broken again?

Social withdrawal and loss

Withdrawal need not occur only in response to bereavement. Social withdrawal and subsequent isolation coupled with a paradoxical dependency are almost general features of ageing. It is here that the family structure is most important in countering the unwanted and unpleasant effects of social and emotional withdrawal by an old person. Of course this can be seen to be largely self-protective, a defence against encountering an unpleasant surprise or yet another personal loss. When we are obliged to live alone, when we have no choice in the matter, we often find it to be a very unpleasant experience. There are no joys when solitary living is forced upon us.

The family remains the bastion from which an old person

may negotiate the social adjustment required to withstand the misfortunes of bereavement, physical illness, and loss of function and status. Many countries maintain their mourning customs, traditional practices to support the bereaved and encourage acceptable ways of expressing grief. Family ties are also maintained to this purpose, although to a much less obvious extent in developed countries. When family structures are large and there are many members in each generation, spending time informally with older generations and giving them their special place in family events (weddings, funerals, graduation ceremonies and so on) become less onerous. Unfortunately, modern societies are faced with greatly restricted family sizes. Where only a couple of generations ago families were typically three-generational (grandparents, parents, uncles and aunts, children and their cousins), a modern family may stretch over four or even five generations, but with far fewer representatives in each generation. The lengthening life span of the older family members increases the burden on the younger generations, sometimes intolerably. Figuratively speaking, the shape of families has changed from a low, broad bush with many branches, to a single long stem with a few leaves.

Ageing and alienation

Changes in the position and social integration of old people in society pose a considerable stress on old people who feel less involved and less valued by that society. Some commentators even go so far as to claim that old people are now alienated from the society to which they once contributed so much. Previously, close ties with an extended family and within the local community helped maintain the psychological well-being of old people. Families are now smaller, communities now much more segregated along age strata, and negative attitudes towards old age can cause a deal of distress to old people.

A contrary view argues that old people since time began have always gradually disengaged from their society. Infirmity, repeated bereavements, acquired poverty and reduced opportunities for involvement are the burdens of ageing. Given the opportunity, however, old people do not conform to this caricature. Instead, close observation shows that whilst old people do make fewer social contacts by choice, those that they do seek and maintain are often much more focused and intense than those of younger people. Perhaps old people simply become more economical in their social behaviour, retaining only those social ties to which they can attach obvious value and failing to nurture relationships they gauge will be 'just too much trouble'.

Alienation theory proposes that the five components of alienation – powerlessness, meaninglessness, normlessness, isolation and self-estrangement – are encountered much more frequently as one ages. Fundamental value systems do not go unchallenged in old age, and members of a previous generation can feel completely out of touch with the attitudes and values of younger members of society. These feelings of elderly alienation are linked to a reduced ability to mount a coping response. There are age-related changes in the nervous, endocrine and immune systems that impair the function of the stress response. Some believe that this reduction in coping capacity is the prime cause of increased stress-associated symptoms in old age and increased susceptibility to disease. The failure of old people to mount an effective response to stress can only be understood when biological, social and psychological data are to hand.

Biological rhythms and ageing

The environment contributes to our psychological function in another and perhaps more complex way. Daily routines are driven by the demands of the day. Waking refreshed from sleep, washing, dressing and eating, followed by a familiar journey to work provides the typical start to the working day for most of us. We may manage our day by undertaking complex tasks in the morning when we feel freshest, leaving more humdrum activities until later in the day. For many, the morning is the best time; the afternoon mops up the more repetitive tasks and the routine checking of the morning's work. Home, food and social recreations end the day and prepare us once again for sleep. Of course not every day is the same, and we all differ in how we manage the time we have for work or pleasure or domestic chores.

Someone who works well in the morning and less well in the latter part of the day is known popularly as a 'lark' and those who do better as the day wears on are sometimes called 'owls'. Such differences are widely recognised and help inform us in our choices of working hours and social pursuits. For example, a lark might choose a job with an early start and a busy morning, whereas an owl might prefer to work later in the day, perhaps becoming a regular shift worker. These rhythms in human performance mirror well-understood rhythms in body function.

Throughout the natural world, rhythms in biological and psychological functions are central to the integration of animals into their environment, especially in anticipation of changes in food supply, extra energy demands such as migration, and the requirements of the breeding cycle. The time cues that keep these rhythms together are taken from the environment. In lower animals the key ability is discrimination between long and short days. This in turn is based on the ability of the eye to receive light and send a message to the central time-keeping cells in the brain. These are clustered together just above where the main optic nerves cross the 'suprachiasmatic nucleus'. Higher animals such as humans maintain more complex biological and psychological rhythms. These run in step with the twenty-four-hour day, changing gradually between the longest midsummer day and the shortest day in midwinter (hence their description as *circadian* – 'about a day'). If we are placed in conditions of temporal and social isolation we cease to keep in step (entrained) with our environment; instead, our rhythms begin to 'free run'. Entrained rhythms keep to a twenty-four-hour cycle, but in free running conditions the cycle lengthens to about twenty-five hours.

Our circadian rhythm follows the daylight length, the social cues that are important to us, the taking of regular meals, and our work schedule. It keeps us in step with one another and probably helps us function at our maximum efficiency. These rhythms characteristically anticipate the demands of the day. The role of the stress-response hormone cortisol is of interest here. The secretory pattern of this hormone shows a regular twenty-four-hour rhythm that is largely independent of exposure to stress. Cortisol is secreted in a pulsatile fashion, with about seven to ten bursts of secretion every day. Most of these cortisol pulses take place between 0400 and 1200 hours, after a period of at least six hours throughout the night with no cortisol bursts at all.

Why should we start secreting cortisol – a stress-responsive hormone – at four in the morning when most of us are asleep? The best guess is that we secrete cortisol in anticipation of the day. It may be that our ancestors who spent perhaps a hundred thousand years as hunter-gatherers coped better with the early morning if they increased baseline cortisol secretion before encountering stressors. Certainly, large daytime predators hunt very actively at dawn.

Another hormonal relic from our prehistoric past is the circadian variation in urine production. Night is a hazardous time for species adapted to daylight living, who might fall victim to nocturnal predators if obliged to rise from sleep to pass urine. Our kidneys have adapted to this problem by slowing the rate of urine production overnight. In old people, control of this rhythm is gradually lost. The fall and the almost inevitable fractured femur incurred while navigating a loose stair carpet by the light of a forty-watt bulb is a direct consequence not of failing kidney or bladder function, but of the loss of the circadian rhythm in urine production.

When the natural synchrony is lost between rhythms within the individual and the environment, psychological function suffers. Good examples are provided by rapid long-distance (transmeridian) travel and shift work, which are notoriously disruptive to well-being. Studies of psychological performance show test scores are worse after such disruption in psychological rhythms. Eventually the rhythms fall back into step and test results normalise gradually after the period of disruption.

Old people maintain their internal rhythms only with difficulty. First, since we are largely social beings and rely on social cues to keep us 'entrained', we must remain in some sort of social framework that will reinforce the cues provided by day length and regular meals. The processes of social disengagement, poor-quality sleep and gradual sensory loss conspire against the maintenance of internal synchronisation of rhythms. For some people this disruption of circadian organisation is sufficient to explain the gradual decline in cognitive performance of old age.

Ageing and sleep

Poor sleep is amongst the most common complaints of old age. Old people say that they sleep badly, have difficulty falling asleep and staying asleep, and wake unrefreshed. At first thought, there could be a myriad causes to reduce the quality of sleep in old people. Changes in bladder function or pain discomfort from physical disease are obvious possible reasons. Poor mental health, persistent anxiety or depression or even dementia could disrupt sleep. Almost fifty years of sleep research has added a new dimension to fairly common-sense understanding. Anyone who has nursed premature infants knows how long they can spend fast asleep. A newborn premature baby who is comfortable will spend almost twenty-four hours continuously sleeping. Later, as awareness of hunger and thirst is acquired, the baby will wake to feed, returning quickly to sleep. Across the life span, the time spent asleep gradually reduces. There are numerous observations from literature. For example, Herman Melville observed in his novel *Moby Dick* that 'old age is always wakeful; as if, the longer linked with life, the less man has to do with aught that looks like death.' Dr George Sigmond had earlier written in the medical journal *The Lancet* that 'the duration of sleep should be, in manhood, about the fourth or the sixth of the twenty four hours; children, the younger they are the more sleep they require; in advanced age there is more watchfulness.'

In modern times, the old are more commonly prescribed hypnotic drugs to induce sleep than are younger adults. For almost thirty years, almost a million old people in the UK have taken a hypnotic before going to bed almost every night, and old people are now the most frequent and prolonged users of these drugs. Prescriptions are not issued on this scale without a huge demand.

Do old people sleep badly – or do they just think they do?

Sleep is regulated by the brain. Brain mechanisms that regulate sleep are affected by ageing. This has been well established by sleep measurement techniques such as sleep encephalography. In this method electrodes are attached to the scalp to detect the electrical activity of the underlying brain areas. The electrodes record voltage differences of a specific amplitude that change with characteristic frequencies, appearing as regularly spaced

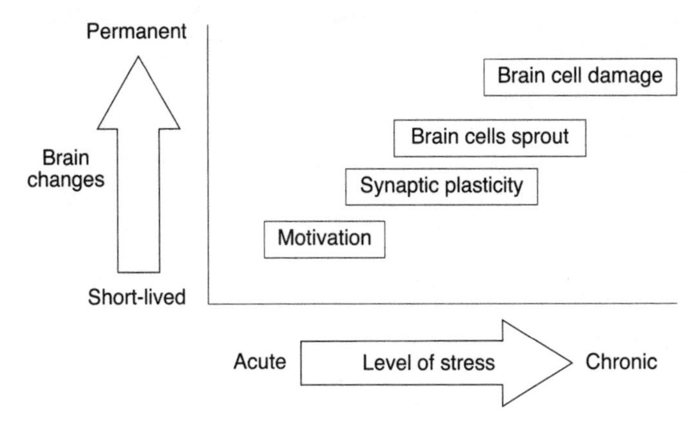

Figure 10 **Stress and the brain**. The figure shows the link between the level of stress and the duration of brain changes induced by stress. The ageing brain has fewer reserves to cope with stresses which tend to produce more enduring damage and accelerate other age-related brain changes.

waves. These waves are the key to measuring and understanding the electroencephalogram (EEG). In the 1950s Eugene Aserinsky and Nathaniel Kleitman observed that regular periods of rapid eye movement (REM) occur during sleep, and opened the door to a path of discovery of changes in sleep pattern and length in health and disease. These patterns show typical changes after consuming alcohol or drugs, and with advancing age. The technique of sleep EEG recording reveals that during sleep old people are more likely to wake than younger people and when they do so they remain awake for longer periods. Unwanted wakening from sleep is an annoyance that when protracted can leave the old person with the impression that the night's sleep has been particularly poor and unrefreshing.

The sleep EEG also reveals that the total amount of time spent asleep lessens gradually with age, from an average of seven and three-quarter hours in young adults down to six hours or less amongst those aged over fifty. These scientific findings confirm what old people have been saying for centuries: you sleep less well as you get older. Of course, some old people can spend a fair amount of time catching a nap, especially in the afternoons. Even when the time spent in daytime naps is taken into consideration, old people spend less time asleep than do young people.

The same is also true for the depth of sleep. The experiences of light and deep sleep are familiar to most of us. These have their counterpart in the sleep EEG which provides evidence of the so-called 'architecture of sleep': drowsiness (stage 1), light sleep (stage 2) and deep sleep (stages 3 and 4). During deep sleep, the rhythms of electrical change across the surface of the brain gradually slow down (slow-wave sleep). Old people spend much less time than youngsters in deep (slow-wave) sleep and proportionally more time in light sleep. This last feature is related to the fact that old people are also more easily roused from sleep. A slight noise can waken an old person, even one who is hard of hearing.

The causes of poor sleep in old people remain something of a mystery. The first level of explanation could be that poor sleep can occur at all ages and that poor sleepers are more likely to survive into old age. This seems unlikely. Another explanation is that the same factors that cause poor sleep in young people also cause poor sleep in old people. Mental health problems such as anxiety or depression are well known to disrupt sleep.

With advancing age, the problem of poor sleep becomes increasingly complex. Poor physical health, acquired mental health problems, difficult housing conditions and ageing brain changes each contribute in their own way to the diminishing ability to fall asleep, remain asleep, and wake feeling restored and refreshed.

Anxiety disorders

Agoraphobia is an irrational fear of open spaces. It was first described by Westphal in 1871, and his first three cases were all men. Some years later, Legrand du Saule described a woman with this condition, but stressed that this was unusual as most cases were male. This sex ratio has been almost completely reversed in modern societies, where women consistently outnumber men in surveys of agoraphobia. The notable exception is Saudi Arabia, where men patients far outnumber women. Why should the sex ratio reverse so completely, and why should Saudi women be so different? In nineteenth-century Europe only the affluent could afford to consult a physician for treatment of mental symptoms, and respectable women of this class never left their homes alone. In modern Saudi Arabia the same is true; Saudi women rarely venture outdoors without a male relative in attendance, and it remains illegal for them to drive cars. If you never go out alone, you are unlikely to exhibit the signs of agoraphobia.

The position of old people in contemporary society is somewhat similar. Many can give good reasons why they should not go out alone: the need to be near a toilet (because of bowel or bladder problems), cardiorespiratory discomfort on walking, and so on. If old people successfully avoid ever going outdoors then they will, like upper-class Victorian ladies, never experience agoraphobic symptoms. The problem of detection of agoraphobia illustrates the difficulties of recognising and understanding anxiety disorders in old people. If a disorder does not present for treatment because the sufferer avoids exposure to whatever triggers the symptoms, some might argue that it is of little interest to health services. The reality is that many old people – estimates from community surveys suggest at least 10 per cent – describe symptoms attributable to anxiety. The severity of these symptoms is sometimes difficult to assess, but they are probably sufficient to impoverish many aspects of social life outside the immediate family. Anxiety symptoms in old people do have economic costs. An inability to use public transport requires private or subsidised travel to make even the simplest journey. Costs of hospital attendance, shopping trips and visits to friends may all contribute to a reduction in available income, or impose additional costs on the health and social security budget.

Old people who suffer from agoraphobia usually develop the symptoms for the first time in old age. So-called 'simple' phobias (fear of spiders, for example) are usually present from childhood and cause little social impairment. By old age, most people have learnt how to adjust to them. Agoraphobia in old people is quite different. It causes moderate to severe impairment in two out of three cases. The disorder may be superimposed upon a long-standing simple phobia or, more commonly, follow an unhappy event such as a fall out of doors or actual trauma such as an accident or street theft. Recovery from a period of physical illness in old people may be protracted and incomplete. These prolonged episodes of minor physical disability are sometimes linked to a failure to regain confidence, so that 'resting at home' slowly blurs into an unwillingness to venture out of doors. This presents a very real problem for family and carers. Obviously, no one wishes to cause distress by forcing an old person to undertake a minor errand alone and unaided. But if such competence remains unrestored, the prospects for independent living can seem very slight indeed. A vicious circle may develop in which concerned families fear for the physical health of the old person. Slowly, the housebound person becomes increasingly disabled by lack of exercise, muscle power ebbs, and cardiorespiratory reserve declines; and a life which should be rewarding and involved in old age becomes an impoverished repertoire of routine visits from paid carers, and diminished involvement in social and family affairs.

Depression in old age

To be old and sad seems a not unreasonable condition for the last stage of human development. Friends, lovers are dead. Children are frightened by your obvious mortality, knowing how much they will miss you – or, more correctly, the person you once were. Your children are now your carers. 'At last,' said one not-so-old lady of her neglectful, absent offspring, 'I'm old enough to be a problem to my children!'

Psychiatrists know depression often follows a stressful event, as night follows day. It is never so clear that the link is causal. People about to become clinically depressed may not be functioning as well as usual. At the edge of competence in managing their affairs, they may not cope with the unexpected as well as they once would have done. Friendships seem fewer, too, and the friends who survive have troubles of their own. This is probably more true of old people in large cities than it is in less densely populated areas. Families grow used to coping with fewer material resources in the face of rural poverty, compared with cities where even short journeys are burdensome and families become dislocated by pressures of urban life.

Causal models of depression – whatever the age of onset – always include roles for involuntary social isolation, lack of social support, recent stressors, loss of a confiding relationship and physical illness or disabilities. These factors arrive as relentless certainties towards the end of life. Compounded by sensory deficits, reduced mobility and declining mental ability, it is a surprise that so many old people cope as well as they do. The spirit of old age is not to give in to the inevitable decline. To do so is to accept that life is over and that there is simply nothing left to give.

The fact is that many old people do not succumb to depression in old age. Their resilience in the face of adversity is intensely remarkable. Questions trigger trite answers, but a little time spent listening to someone who has aged successfully, who is not demoralised by the reality of old age, can be especially rewarding. Erik Erikson coined the dilemma 'ego integrity versus despair' and summarised the predicament for many old people. Other writers have observed the 'inner peace' achieved by some old people, especially in their ninth and tenth decades of life. How is this done? What are the preconditions necessary to achieve this successful adjustment at the extreme of life?

Typical responses in an irreligious age demand examination of modern equivalents of terms such as 'wisdom' and 'spirituality'. The words vary between cultures, but the 'successful ager' reflects on life lived with its accumulation of successes and failures, some prizes, some penalties, and on a death foreseen. The outer serenity of these old people may be deceptive. Listen to what they say, and you will hear passionate concerns about the world and what will become of us. Their success is probably more to do with their continuing involvement with the larger world than anything else. For sure, we do not know. Success must in most part depend on keeping hold of the place – not losing the plot – on the retention of mental abilities. In Erikson's terms, ego integrity relies on these abilities more than on anything else.

Depression and demoralisation in old age

Depressive illness in old age is not just the demoralisation of ageing. It is quite distinct from that. Depressed old people suffer a pervasive sense of unhappiness which affects all that they do. A turn for the better does not cure their sadness. The symptoms are well known: hopeless preoccupations with the inevitability of their pointless, meaningless life; their burden on society. Nothing, the old depressed person is convinced, can be done to change what is sure to come about. Concentration is poor, memory seems to be failing: 'All I can think about is what I've done wrong, how I've let the people I loved down so much that I don't deserve what they are stupidly trying to do for me now.' Sleep and appetite are disrupted, trivial worries dominate everyday thinking and distortions of reason abound. Even with money in the bank and a regular pension, these worries may assume monster proportions. 'There's no money for the bills ... I can't afford to eat ... I'm sure I have an awful disease – cancer or something.'

Meantime, money goes unspent, although the cat is well fed, and the terrible crime, sin or disease is simply explained by an oversight, a gentle flirtation some forty years ago, or something as insignificant as an episode of constipation.

Delusions and hallucinations in old age

Cognitive impairment alone is not the major cause of hospital or nursing-home care for an old person with failing mental powers. Delusions and hallucinations, and the corresponding behaviours, most often trigger the call for urgent help.

Delusions are a way of reconstructing the external world which takes account of the emotions or uncertainties experienced by the deluded person. Delusions always contain an element of falsehood or illogicality, are always held with firm conviction, and are not understandable in terms of the widely held religious or cultural beliefs of the day. Application of definitions such as this to the accounts offered by some old people to explain their behaviour is by no means easy. Who is to say that a relative is not stealing money, or that uninvited strangers are not entering the homes of old people who have kept trust with the values of the older generation?

Case 2

Cathy B. brought her seventy-nine-year-old mother along to the clinic. She said her mother had become forgetful some years ago, she couldn't remember precisely when. Her mother was able to shop and sometimes visited old friends, but had in recent months begun to complain of strangers in her house. She said she knew that strangers had been in the house because some of her belongings had been stolen. Later, her daughter found the 'stolen' items apparently mislaid in her mother's home. At first the family were amused by the complaint, as it always concerned trivial items of no likely interest to a thief. ('Grandma,' she had been told, 'Nobody would break into your house to steal a family photograph album!') However, the complaints became increasingly tiresome to the daughter when a neighbour – whose son was at one point accused – and a genial community policeman became involved. The event that triggered the referral was bizarre. One evening the mother knocked angrily at her daughter's door having walked the half-mile or so from her own home. She demanded that her daughter immediately return with her because the thief had been caught in her house. In fact, as she spoke, she was certain he was there fighting with policemen. Suddenly she yelled, 'Look, they're here too!' and pointed to the television. Her daughter realised that her mother had mistaken a police drama series for real events in her own home. She gently quietened her, reassured her and then asked if she had remembered to lock her house before leaving. 'Oh, yes,' said her mother, 'here are the keys' – and handed her daughter the remote control for the television.

In this case the daughter is apparently doing all she can to maintain her mother's home – the family house in which she had grown up and where family reunions were often held. However, the daughter's efforts came at increasing cost to herself. The delusion of theft of property is commonplace in old people and may be related to their need to explain the inability to locate a missing item. It is particularly troublesome when accusations, sometimes made with appalling rudeness, are made by an old person whose only hope of retaining a measure of personal independence is the continuing support of one of their children (usually a daughter). The misidentification of television actors as real individuals is sometimes linked to other misidentifications which can even include the failure to recognise one's own reflection in a mirror.

Numerous theories are put forward to explain why old people are more likely to develop delusions. Although cognitive impairment is a well-established risk factor for delusions, it is not established as a prerequisite. The fact that delusions so often arise in the setting of cognitive impairment suggests a number of possibilities. In one, cognitive impairment alone is proposed to elevate flawed reasoning to a dominant position in mental life. Without the cognitive strategies to test the accuracy of an argument, a false belief may be seized upon as the obvious and sometimes the only possible explanation. This does not explain why in so many instances the delusions that develop in old people become firmly held and enduring, not just for a few hours at a time, but for days or even months. A second possibility may be more plausible. Perhaps the disease mechanisms that cause memory impairment in old age can also cause delusions to arise and persist.

A small group of people develop severe mental illness for the first time in the seventh, eighth or even ninth decades of life. Left untreated, these disorders persist until death. The delusions and hallucinations are typically persecutory in nature. The term 'paraphrenia' (introduced to describe something that is 'partly schizophrenic') was used in the early part of the twentieth century. The term fell into disuse until restored to life by Martin Roth in 1955 to describe the 10 per cent of old people first admitted to a mental asylum after the age of sixty years but with no features of dementia. At that time, British psychiatrists placed great stock on the diagnostic process and believed that the final diagnosis conveyed a meaningful scientific understanding of the disease process blamed for the mental symptoms. Such reasoning is highly suspect, and was recognised as such. The exception proves to be that group of diagnoses where the disease mechanism is directly observable from study of brain tissue. An obvious example would be the mental symptoms following stroke, where a demonstrable brain lesion precedes the first mental symptoms and there is other evidence to link the site and extent of the lesion with the nature and severity of the symptoms. This approach to diagnosis does not rest entirely on the patient's own descriptions and, critically, it is often possible to establish the principle of 'temporal precedence'. Here, the causal event occurs before symptom onset and produces observable change in bodily tissue involved in symptom production.

The diagnosis of paraphrenia never caught on in the USA. There, psychiatrists preferred the label 'organic brain syndrome' if there were demonstrable brain pathology, 'schizophrenia' if the symptoms merited the diagnosis, or 'persistent delusional disorder' if hallucinations typical of schizophrenia were absent. For American psychiatrists, the principal diagnostic questions concerned the detection of brain disease and how widely the diagnosis of schizophrenia could be applied. The nature of hallucinations in old age can differ markedly from those encountered in young people. In the absence of cognitive impairment, old people will exercise caution in their interpretations of hallucination. To a variable extent, they retain the ability to judge whether a sensory stimulus arises in the external world or is a product of their internal 'thinking' space. This is particularly relevant when hallucinations occur during a period of normal grieving. It is not uncommon for an old person, bereaved after decades of marriage, to have auditory, visual and sometimes tactile experiences attributable to the presence of the dead person. These experiences are rarely fearful and can be comforting. They can also be linked to delusions that the dead person continues to frequent the home, requiring changes of clothing and meals at regular times.

Case 3

Mary C. had grieved for her husband's death for about fifteen months before psychiatric opinion was sought. When seen at home she was found to have a moderate degree of cognitive impairment suggestive of a dementing process of about two years' duration. She spoke of her husband as though he were still alive, pointing to the place she had set for him at table and confidently asserting he would soon return from work and eat the meal she had begun to prepare. Enquiry elsewhere revealed her husband to have been a well-known and somewhat charismatic local professional singer who had 'swept her off her feet as a girl before the War'. After a brief courtship she married him but remained childless. Friends remembered the remarkable strength of her devotion to him and her insistence on his returning home from social or professional engagements in good time for a meal, an early night and so on. He had acquired a reputation (before and after marriage) of being a charming rogue, with 'an eye for the ladies'. Her friends thought that her domestic strategy was intended to set limits on his behaviour, and they speculated that even after death his widow continued strenuously to keep him away from other women.

On examination Mrs C. showed memory impairment, but not to such an extent that she could not manage alone at home. Her more striking disability was caused by bilateral cataracts which much impaired her vision.

Sensory impairment is major risk factor for delusions and hallucinations in old age. It is such a frequent 'bystander' that much thought has been given to its possible causal role. Association between phenomena does not prove that one causes the other, only that they are associated in time. However, it is tempting to speculate why so many deluded and hallucinating old people suffer some degree of sensory impairment. A possible explanation is that in undamaged sensory perception, the continuous acquisition, organisation and understanding of sensory stimuli saturates the capacity for reactivation of stored memories. This type of reasoning rests on the view that memory traces are stored in specific brain areas that also serve to detect and make meaningful the sensory experience. In these terms, memory is modality-specific: for example, visual memories will be stored in the visual part of the brain, auditory ones in the auditory part, and so on. When the sensory input falls below a critical threshold level and consciousness remains unimpaired, there will be a spontaneous release of stored memory traces which emerge in consciousness as perceptions. It is difficult to find empirical support for this proposition. At one level it seems plausible that reduction of sensory input might facilitate the recall of images. This is certainly true in experiments where sensory input has been either reduced or grossly exaggerated. In the reduced setting, subjects may more easily recollect and describe a scene from memory, for example using visual imagery, with or without eyes closed. Likewise, 'white noise' played continuously over a long period can induce hallucinatory experiences in the absence of cognitive impairment.

Psychiatrists have learned a great deal from neurologists about the mechanisms in the brain involved with verbal hallucinations. Enfield and his co-workers in Montreal conducted a series of experiments to show that stimulation of the parts of the brain involved in auditory processing could result in crude 'noises'. Other (secondary) auditory association areas of the brain produced more complex sounds when stimulated. Complex verbal hallucinations (voices) are much more difficult to reproduce. Those described seem qualitatively and quantitatively to be distinct from the typical verbal hallucinations.

Survival advantages and mental ability

Retention of mental ability in late life improves our chances of living longer. Even in the face of failing physical health, most research shows that old people who maintain mental skills and a good memory enjoy a better quality of life than those who do not. The opposite is equally true: mental decline in old age is a powerful predictor of an earlier death.

This link between mental ability and late-life survival is not simply explained by the gradually progressive nature of a dementia like Alzheimer's disease which causes premature death. More probably it is due to several factors acting in concert. One of these is almost certain to be age-related brain blood vessel disease, which increases the chances of a stroke or a fall, with often fatal consequences. Less obvious is the fact that mental decline is symptomatic of a failing brain functioning at 'or very close to' the limit of its ability. In these circumstances, the ageing brain fails to manage effectively the complex array of tasks needed to regulate the body's internal environment, anticipate regular changes in the outside world, and orchestrate responses to challenges from within and outside the body. The capacity to do all these things can be summarised as 'integrity of brain function' and is essential for life. Minor impairments of brain integrity may not necessarily be fatal, but will cause lasting disabilities and sometimes early death.

Who is at risk of brain ageing?

This question seems to have an obvious answer: anyone who is old. But it is not so simple. In the 1950s and 1960s communities were identified in which people maintained physical performance levels into late middle-age despite the sort of diet that is considered to have potentially disastrous consequences for heart disease. Individuals of the same calendar age do not invariably age at the same rate.

The best information about differences between individuals in brain ageing comes from the studies on stroke and dementia discussed in the following chapters. These studies point to the first consistent finding: individuals who have lower-paid jobs, lower educational attainments, a 'poor' diet (less fruit and vegetables) and who smoke are at greatest risk. Of course, these factors are not unrelated to each other. They are each components of an unhealthy lifestyle linked repeatedly to chronic diseases and premature death. Elegant mathematical models of these factors and their associated diseases point to substantial interactions between the risk factors, and the likelihood that their effects on the brain are cumulative and may hasten the effects of ageing. An alternative view is possible, however. Maybe there are 'compensatory' or 'protective' factors in healthier lifestyles, so that the consequences of the 'unhealthy' lifestyle are the expected associations between all the trials and tribulations of life and diseases such as stroke and dementia. In these terms, the healthier lifestyle may reduce risk. How would that come about?

The answers to these questions are unknown. Current research relies on clues from studies of differences between healthy old people and those who have more of the diseases linked to ageing but are of the same calendar age. The first studies relied on historical information almost inadvertently retained about the health of old people when they were infants or even about their mother's pregnancy. The results, which remain controversial, suggest links between maternal health, birthweight and vascular disease in middle-age and late life. The best studies of this type are prospective and long-term, and from time to time information about the source of these differences is published and new ideas developed. Events in early life are of some (but not great) importance and may increase our risk of late-life vascular disease including risk of stroke. Low childhood educational attainments and mental ability increase the risk of late-life dementia.

The link does not depend on well-known associations between poor home circumstances or chronic childhood diseases. Some of the risk is likely to be genetic but adult environmental factors are certainly important. The current studies point to lifestyle factors as having most relevance. When the health of middle-aged people is carefully followed from the age of fifty or so into late life (up to eighty years), those who maintain mental ability best are those who smoke or drink least (if at all), eat a balanced diet with fresh fruit, vegetables and fish, and keep physically active. There is some evidence that maintenance of mental activity reduces the risk of mental decline, but this is by no means certain. What seems likely is that all these components add up to a healthy lifestyle which significantly reduces the risk of common chronic diseases. In turn, mental functions are less affected by poor health, with all the well-known consequences for self-esteem, sense of purpose and involvement with others.

Chapter 7

When blood stops

Brain blood flow, stroke and ageing

Every cell in our body needs blood. When blood flow stops the cell will enter a critical state, and if the flow is not restored this critical state will trigger a cascade of events that will – usually within minutes – kill the cell. Although some cells have alternative sources of energy that do not require oxygen, brain cells do not; they rely on a constant supply of oxygen to convert their primary energy source (sugar) to brain cell energy. Without oxygen and the removal of cellular waste material, the brain cell is dead within ten minutes of a stoppage of blood flow. The heart and brain must be protected against such tissue death if at all possible, and so they both enjoy certain privileges when the central blood supply is compromised. The brain will be protected at the expense of other organs such as the gut and the muscles. As far as possible blood flow to the head will be preserved.

The brain is also protected from harm in a way uniquely distinct from other organs. The linings of the blood vessels in the brain are made up of cells that are tightly linked to one another. There are few ways to enter the brain from inside the blood vessel other than to pass through this lining, which is called the 'blood–brain barrier'. It can be imagined as a continuous sheet composed of a single layer of tightly fitting cells, rather like a patchwork quilt with the pieces tightly sewn together. Like all biological membranes, the cell walls of the blood–brain barrier have outer protein layers and an internal fat (fluid) layer. Some small molecules can dissolve in the fat layer and enter the brain by the process of diffusion. Larger molecules rely on active transport mechanisms to pass through the membrane. These active transport mechanisms are large membrane-spanning molecules that function rather like revolving doors at the entrance to a large store. On the outside (facing the inside of the brain blood vessel) they combine with a molecule looking for a way into the brain. The active transport molecule then carries the substance from the blood vessel across the blood–brain barrier and releases it on the inside face of the membrane, allowing the passenger molecule to move about the brain. These active transport mechanisms are highly selective and help protect the brain from unwanted molecules (for example toxins) which may be produced elsewhere in the body or consumed with food.

Blood flow through the brain deteriorates slowly with age. Imaging techniques such as positron emission tomography (PET) show this very well. The decline occurs in all brain areas and is matched by a similar drop in the use of oxygen by the brain. Brain blood flow is tightly coupled to brain work (brain metabolism). At first glance the brain seems to work less strenuously in old age and so needs less blood. This may not be true. Decreased brain blood flow could be caused by other factors. First, if ageing brain cells tended to die (and we know that they are not often replaced), less blood would be needed. Careful imaging studies show that the brain gradually shrinks with age (perhaps by about 10 per cent between the ages of twenty and sixty years) but the number of brain cells does not decrease so dramatically. The case that the decline in brain blood flow with age simply reflects decreased brain cell number is not strong.

If brain blood vessels were affected by disease, might this be sufficient to impede blood flow through the brain? While there is some evidence for this, it is unsafe to jump to the conclusion that brain blood vessel disease is caused by ageing. Certainly, it is more common with ageing (that is, it is 'age-related)', but it is not 'age-dependent'; there are too many examples of individuals living into their ninth, tenth or even eleventh decades of life with no evidence of brain blood vessel disease. A third possibility is that there may be functional changes in the regulation of brain blood vessel diameter (vasomotor reactivity). In health, the brain is protected from extremes of high or low blood pressure; high pressure might damage the blood vessels of the brain, and low pressure might diminish blood flow to the point that respiration in brain cells would be compromised. The mechanism ensuring that brain blood flow is reasonably constant in the face of a wide variation in blood pressures is called 'cerebral autoregulation', meaning that the brain itself controls cerebral blood flow. The brain does this by changing the diameter of the blood vessels, contracting the muscles around the vessel when pressure is low (increasing flow) or relaxing the muscles when pressure is high (decreasing flow). These contractions take place in medium-sized blood vessels and are highly efficient. They ensure, for example, that during the large and rapid changes in blood pressure accompanying normal daily activity we do not lose consciousness or burst a blood vessel. There is a fine network of nerves from the brain to the muscles around the brain blood vessels. These nerve endings use a wide range of chemical transmitters which can oppose each other in quite delicate ways, allowing fine control of cerebral blood vessel diameter. Acetylcholine can dilate blood vessels in the cortex, whereas serotonin can constrict large vessels in the brain stem or dilate the very smallest blood vessels in the brain.

Old people with high blood pressure (hypertension) are commonly treated with pressure-lowering (antihypertensive) drugs. Some of the older antihypertensive drugs can disrupt autoregulation and cause many side effects. Newer drugs are tailored to avoid these side effects and allow gradual readjustment of autoregulatory mechanisms while lowering overall blood pressure. Hypertension, however, is not the only culprit: diabetes, raised blood fat levels (hyperlipidaemia), heart disease and cigarette smoking all reduce brain blood flow. The good news is that regular physical exercise (including paid employment) is linked to maintenance of normal cerebral blood flow during ageing. Old people who continue to work after retirement or continue regular physical exercise show much more constant cerebral blood flow values during ageing than people who retire and engage in little if any physical activity. Increased brain blood flow has been linked to maintenance of good scores on psychological tests. Some element of self-selection could be at work here: maybe those who are able to maintain cerebral blood flow are also able to go on working, whereas those who cannot find work too effortful. Brain work will increase blood flow to that part of the brain doing the work. Maybe, for these mechanisms to remain efficient and intact, they must be constantly used. Perhaps they become inefficient with disuse.

It is difficult to measure blood–brain barrier functions in humans. Large molecules that normally cannot enter the brain (because there is no active transport mechanism) can be found in increased concentrations in brain during ageing. This suggests that the blood–brain barrier becomes 'leaky' with time, although it is never clear if this is a consequence of ageing alone or relies on an age-related disease process such as cerebrovascular disease. Once cerebrovascular disease is established the normally smooth flow of blood through the vessels becomes turbulent, with pools and eddies replacing the straight flow in the small vessels. Eventually, damage is caused to the delicate inner lining of the blood vessel, compromising the blood–brain barrier and causing the delivery of essential nutrients and removal of waste products from the brain to become progressively less efficient.

Structural changes can also be seen in the wall of the blood vessel. In healthy adult blood vessels the amounts of the protein collagen (which provides stiffness) and elastin (which provides elasticity) are balanced, giving a flexible yet supportive vessel wall. With ageing increased amounts of collagen make the blood vessel less distensible and consequently less able to narrow or dilate in response to changes in blood pressure. This 'stiffening' is not a sudden change with ageing, but begins slowly from about the age of twenty-five years.

Vascular dementia

In 1968 dementia researchers met in London to talk about the causes of dementia. This was a historic meeting. Until that time cerebrovascular disease was firmly believed by both doctors and members of the public to be the most frequent cause of dementia in old age. 'Hardening of the arteries of the brain' was thought to be the most common cause of mental decline in old age, whereas Alzheimer's disease was believed to be a rare dementia in middle age, and was little taught to medical students. A group of researchers in Newcastle-upon-Tyne triggered a reversal in thinking about the likely causes of dementia and what might be done to prevent them.

A massive collaborative international effort was soon under way to investigate Alzheimer's disease, while vascular dementia became relatively neglected. In fact, if anything Alzheimer's disease was overdiagnosed. The end of the twentieth century has seen a reawakening of interest in cerebrovascular molecular pathology, the probable cause of most cases of vascular dementia. Much to the surprise of clinicians who had made careful efforts to distinguish between Alzheimer's disease and vascular dementia, the two disease processes were now seen to be closely linked.

Before one can study a disease process, it is necessary to define the characteristics that distinguish it from other disorders. The diagnostic process not only aids research but is also essential to understand what might happen to the patient (the prognosis), to assess risks to other family members (by genetic analysis) and to seek to reduce harm (by treatment or removal of risk factors). The problem with studying vascular dementia is that it is actually a group of several overlapping disorders. The most common type is caused by brain infarcts (an infarct is an area of dead brain tissue killed by failure of blood supply), usually due to blockage of the blood vessel supplying that area, but sometimes due to a burst in the vessel wall (usually at a spot weakened by disease or congenital abnormality) so that the blood floods adjacent brain tissue. In multi-infarct dementia there are numerous infarcts in the cerebral cortex and occasionally elsewhere. Sometimes, however, vascular dementia is caused by a single infarct in a part of the brain critical for memory (a 'bottleneck' structure).

There are some uncommon causes of vascular dementia, of which the best-known is probably Binswanger's disease. This is linked to loss of the myelin coat of the nerve tracts in the brain white matter, and multiple areas of brain cell loss. Because these look like little empty pools when seen down a microscope, they are termed 'lacunar infarcts'. Cortical infarcts are also linked to another rare type of dementia which has proved to be very informative about the genetic causes of vascular dementia. Other types of vascular dementia affect mostly the smallest blood vessels in the brain.

The most common type of vascular dementia (multi-infarct dementia) can be found in the brains of between one in ten and one in five of all elderly people with dementia in developed countries. There are striking differences in the relative proportions of vascular dementia and Alzheimer's disease between northern Europe, central and western Africa, and Japan. In northern Europe Alzheimer's disease is the most common type of dementia (about one in two), whereas in Japan vascular dementia is more common (about two in three), and in black Africans, vascular dementia predominates and Alzheimer's disease is uncommon – even rare.

The clinical features of vascular dementia are usually quite different from Alzheimer's disease. The diagnosis is easiest to make when the dementia follows a stroke. In fact, most diagnostic systems prompt a diagnosis of vascular dementia when a stroke has occurred. This does not make the diagnosis of vascular dementia impossible when there is no stroke, but it is more difficult. The main differences between vascular dementia and Alzheimer's disease are as follows. In vascular dementia the onset is usually sudden, whereas in Alzheimer's disease the onset is gradual. The progress of vascular dementia follows a stepwise course, in which sudden deterioration is followed by a slight improvement, a period of stability, a further episode of deterioration, and so on. In contrast, Alzheimer's disease usually progresses gradually. In vascular dementia personality and judgement are reasonably well preserved in the early phase of the disorder, but depressive symptoms, emotional volatility and episodes of prolonged crying or laughter for no appropriate reason are not uncommon. There is also a tendency to become confused at night (probably because of disturbance to the cerebral blood flow when lying down).

Performance on psychological tests is usually unhelpful in the diagnosis of vascular dementia. In Alzheimer's disease there are some typical psychological features such as specific types of hand–eye control and certain forms of language difficulty. These are described further in the next chapter. A slowly progressive dementia without neurological deficits is rarely attributed to vascular disease when the brain is examined after death.

The diagnosis of Binswanger's disease is much more difficult to make during life. It is an uncommon cause of dementia and most specialists see very few cases in whom they can make the diagnosis with confidence. Almost all patients with Binswanger's disease have a long history of hypertension. They usually show a slowly progressive dementia and a history of numerous small strokes. Impaired speech, walking and balance are amongst the more consistent clinical features.

What is a stroke?

A stroke is a rapid loss of brain function caused by brain blood vessel abnormality. Some people recover within a day, some more slowly, and others can be left severely disabled – or dead. In all countries strokes are slightly more common in men than in women, although the combined total varies considerably. Apart from countries with the highest rate (Russia) and lowest rate (France), incidence rates worldwide are broadly similar – about two first-ever strokes per 1000 adults per annum. The risk of stroke, however, is not evenly distributed within populations. Increased blood pressure, cigarette smoking, irregular contractions of the heart input chambers (atrial fibrillation) and diabetes are all well-established risk factors. The treatment of high blood pressure will reduce the risk of stroke.

As in other areas of brain science, the 1990s saw rapid progress in understanding the basic mechanisms of stroke, and these have pointed to new drug treatments. Stroke is now considered to be a medical emergency equivalent to a heart attack. Sadly, considerable delays in seeking treatment for stroke patients are not uncommon, particularly for old people. In spite of educational campaigns aimed at the public and even some doctors, knowledge that early treatment of stroke is important is not widespread. Doctors who are up-to-date on stroke treatment sometimes tell their colleagues what should happen if they were to have a stroke. All are agreed that urgent assessment is essential by a stroke expert or least somebody with a substantial interest in stroke treatment ('First find an expert!'). If a patient can reach a stroke centre within three hours and the stroke is severe, then urgent brain scan (by computed tomography) followed by assessment for treatment with 'clot-busting' drugs is vital. Not all doctors, however, are agreed that 'clot busters' are necessary for every stroke patient and will reduce the damage done by stroke – some believe these drugs may make that damage worse. However, it is certain that carefully selected patients will often benefit from these drugs. Patients with minor strokes who do not deteriorate should not receive these drugs; in them the risks linked to 'clot busters' certainly outweigh any advantages.

The brain scan will show if the stroke is caused by haemorrhage or, as is more common, by blockage of a blood vessel. If it is a haemorrhage, then drugs that prevent clot formation must be avoided: they will simply increase the risk of more bleeding into the brain. Specialist stroke nursing is valuable early on. Experienced nurses can assess the need for helping a patient pass urine (urinary catheterisation), early mobilisation and oral feeding. The ability to swallow without choking on food or drink needs to be carefully monitored. If stroke rehabilitation is to succeed, patients need to be as well watered and fed as possible. When patients are confidently expected to survive their stroke, oral feeding should always start within the first twenty-four hours.

Strokes tend to recur. Once the patient has regained a reasonable quality of life, low-dose aspirin (75 mg once daily) should be started.

Beyond these three key steps in stroke management (stroke unit care, 'clot-busting' drugs and long-term aspirin), questions remain about stroke management. The most important is probably whether the high risk of long-term mental impairment after a stroke can be reduced. When brain scans reveal multiple areas of brain damage attributable to blood vessel disease, the risk is high of progressive mental impairment. This can occur even without an obvious stroke. The only good evidence so far is that reduction of high blood pressure will reduce the risk of progressive cognitive impairment.

Controlling other risk factors (by stopping smoking, taking more exercise, reducing blood fat levels by drug treatment, and improving control of diabetes) probably helps as well. There is less good evidence to support surgical measures to correct vessel disease in the main neck arteries to the brain (carotid endarterectomy). Correction of abnormal heartbeat (atrial fibrillation), if present, will also reduce the risk of another stroke.

What causes vascular dementia?

There is now good evidence to link longstanding high blood pressure with brain abnormalities detected on brain imaging or after death. These brain abnormalities arise regardless of how successfully the hypertension was treated during life. When other risk factors for stroke are also present it is believed that these increase the likelihood not only of stroke but also of vascular dementia. These risk factors are abnormal glucose metabolism (diabetes mellitus) and hyperlipidaemia. These conditions damage the blood vessel wall and cause atherosclerotic disease. The blockages and bursts in brain blood vessels that follow in turn cause multiple, scattered cerebral infarcts. Cerebral blood flow is reduced in these areas, impairing mental function. Studies of cerebral metabolism in hypertensive subjects using PET scans have identified the watershed area between the anterior and middle cerebral arteries as being particularly susceptible to the effects of hypertension, even when well controlled. The ventricles of the brain are also enlarged compared with people with normal blood pressure. It is as yet unclear whether these brain changes are due to the hypertension or to another disease process that also causes the hypertension. Some researchers believe that there is a disease process that causes hypertension and also progressively damages the brain structures and connections that control blood pressure.

Until the publication of reports by a Canadian neurologist, Vladimir Haschinski, the treatment and prevention of vascular dementia had received little attention in comparison with Alzheimer's disease. For some physicians, vascular dementia had remained the last redoubt of 'essential hypertension'. Here the term 'essential' is used to describe the idea that a diseased organ requires increased blood pressure to maintain its function. In these terms, hypertension was regarded as essential for the organ to function.

This old-fashioned view is certainly wrong. It was supported by clinical experience with the first generation of drugs to control high blood pressure. These drugs have the unfortunate effect of paralysing cerebral autoregulation, leaving the brain unprotected from variations in blood pressure. A sudden drop in blood pressure produced an equivalent drop in cerebral blood flow, and the unfortunate patient could lose consciousness simply by standing up too quickly. When newer, more satisfactory agents were introduced for old people with hypertension and dementia, their benefits were soon established.

A programme of research at the Baylor College of Medicine in

Texas provided the first truly systematic account of the treatment of a large number of patients with vascular dementia. This showed that the progress of dementia attributable to vascular disease could be slowed by reduction of vascular risk factors, the best-established of which are hypertension, heart disease and smoking. The Baylor College group recommended that in vascular dementia it was always worthwhile to seek to slow the disease process by preventing the recurrence of cerebral infarcts; this can be done by controlling hypertension, reducing platelet 'stickiness', lowering blood fat levels if raised, and removing sources outside the brain of embolic stroke (atrial fibrillation and carotid atherosclerosis).

Current treatment

Improvements in the treatment of stroke are relevant to the treatment and possible prevention of vascular dementia. First, improved methods of brain imaging can reveal just where the stroke has occurred, how severe it is, and how reversible are the changes produced in the affected brain tissue. Second, 'clot-busting' drugs such as the older streptokinase and the newer recombinant tissue plasminogen activator can be beneficial in the acute treatment of stroke. Third, the sources of stroke from outside the brain are now better understood, and progress has been made in changing the 'stickiness' of blood constituents and establishing how these might have contributed to the stroke in the first place. These advances are termed 'neuroprotective' because they protect brain cells from stroke; they have an important part to play in the prevention and management of recurrent stroke, and in the treatment and prevention of vascular dementia.

Neuroprotection is designed either to prevent the type of brain injury caused by deprivation of oxygen supply or, if injury has already occurred, to prevent further injury. The inevitable path to neuronal death starts with reduced cerebral blood flow, which in turn rapidly depletes the energy available for brain cells to sustain life. Because oxygen is unavailable, acid accumulates in the cell and causes failure of the brain cell pumps that maintain the mineral salt balance across the extracellular membrane. The membrane rapidly deteriorates, and calcium floods into the body of the cell, activating proteins that will digest key cell components. The only possible outcome is cell death. Meantime, the brain is struggling to deal with depletion in cellular energy stores. Messages of desperation are sent out by the damaged cell demanding immediate reperfusion of the area and calling on inflammatory chemical attractants to bring up any additional help. Unfortunately, these chemical attractants also activate an enzyme that causes reactive oxygen species to be generated; these add to the worries of the cell because they in turn will lead to cell death. Neuroprotective strategies aim to lessen the neurotoxic consequences of lack of oxygen at each step of the cascade.

Chapter 8

Thieves in the night

Dementia and Alzheimer's disease

Some people say they could cope with anything life throws at them – but not dementia. The indignity of mouthing mumbled irrationalities would be too much to bear. To depend on others for everything – spoonfuls of gruel, the toilet, a change of soiled clothes – is intolerable even to imagine.

With support, the first few years of a dementing illness – contrary to many fears – are unexpectedly bearable – or would be if it were not for knowing what comes next. Most dementias of old age are a gradual process, and unless death intervenes, lead inevitably to a complete loss of power of memory, reasoning, and (perhaps hardest of all) even simple conversation.

What is dementia?

Dementia is a mental disorder of adult life. It is most common in old people in whom it is usually caused by Alzheimer's disease, vascular disease (described in Chapter 7), or a mixture of the two. The common forms of dementia are slowly progressive, and eventually all powers of memory and reasoning are lost. The first symptoms reflect the precise cause of dementia and the parts of the ageing brain most affected.

The first psychological features of dementia are the deficits in memory and understanding attributable to impaired brain function; but these are enmeshed with the psychological reactions of the individual to these deficits. Not surprisingly, there are considerable changes in outlook. Emotions in this early phase are coloured by feelings of inexplicable fear, bleakness, a sense of powerlessness or notions of dread, loathing or shame. In moments of despair these feelings are overwhelming and threaten to cause immediate personal disintegration with no hope of recovery. Powers of descriptive language are usually retained in early dementia, and its onset has been graphically described in terms such as 'a visit from a thief in the night'. These words were used by a patient early in the illness to describe the silent, unobserved loss of mental abilities and memories while judgement and insight are retained for a time. Such clarity of expression can make it hard to distinguish between early symptoms of dementia and depressive illness caused by losses typical of old age, such as bereavement.

The first brain pathologists used the light microscope to distinguish between the types of brain disease that caused the loss of mental functions in adult life. These early neuropathologists identified areas of patchy loss of the usual structural organisation of the brain, which did not appear to be associated with a haemorrhage or blockage of a nearby blood vessel. As understanding progressed, careful and systematic accounts of disruption by dementia of the cellular organisation of the brain were reported.

The damage to the brain

Like any other organ, the brain must maintain a minimum level of functional integrity in order to do its work. Chronic disease is associated with chronic symptoms, so that the individual remains at or close to the symptom 'threshold'. The 'functional reserve' of the brain is determined by a number of complex factors such as genetic makeup, childhood illnesses, educational opportunities, diet, and exposure to neurotoxins. We shall consider these factors again when we look at the risks of developing dementia and the problem of 'latency' in the emergence of symptoms.

Alzheimer's disease is associated with the loss of cortical neurons. This loss is well established at death, when typically about 30 per cent of cortical cells are missing. Surviving cells show reduced dendritic sprouting and synaptic formations. The medial temporal lobe often shows more extensive cell loss than other cortical areas, and this loss is much greater than that typical of normal ageing. Genetic factors may modify cognitive decline in ageing, and may combine with unknown environmental agents to accelerate this reduction of the cerebral reserve.

Brain imaging could potentially throw light on these findings, but unfortunately most studies have been of cross-sectional design, reporting group differences between 'demented' and 'non-demented' subjects. Longitudinal studies are needed to reveal the relationship between neuronal loss and symptom development in Alzheimer's disease and, importantly, to address the question of brain size and structure before disease onset. David Smith's team in Oxford (the OPTIMA project) did just this when they studied sixty-one Alzheimer's disease patients, comparing them with forty-seven normally ageing control subjects. Evaluation of brain scans and tests of mental ability, performed annually, allowed the relationship between age and the minimum thickness of the medial temporal lobe to be estimated in both groups. Four patterns of change were observed: one pattern shows little initial alteration in lobe thickness followed by rapid change; in the second (most frequent) pattern there is a rapid rate of change from the first observation onwards; and in the third there is an initially rapid rate that levels out. The fourth type showed no change in thickness over time. There was some suggestion that patients with the thickest medial temporal lobes suffered the mildest forms of dementia. These measurements provide compelling evidence of rapidly progressive changes in the medial temporal lobe, inconsistent with normal ageing, that may reflect a catastrophic event occurring within a specific brain area. This idea is very relevant to understanding what happens to the ageing brain in Alzheimer's disease.

The first neuropathologists saw a brain devastated by Alzheimer's disease and wondered why brain cells should die in such large numbers. If not all brain cells die, but some are spared while others are lost, then further questions arise: is there selective loss of brain cells in Alzheimer's disease, and does this explain the symptoms?

The elegant studies by the Braak team in Germany have demonstrated six stages in the deposition of neurofibrillary tangles in Alzheimer's disease. Their studies have focused on the brain's limbic system and its connections, and suggest an intellectually attractive alternative to the 'cerebral reserve' hypothesis. The limbic system is a complex of brain structures that has passed into and out of fashion over the last century. It is highly relevant to understanding the memory problems of dementia.

Major pathways in the limbic system transfer information from short-term to long-term memory. This transfer requires the brain structures involved to cooperate with one another. Motivation and other emotions work as a team to 'label' the memories as they are stored. This ensures that when memories are retrieved from storage they are linked as precisely as possible to an appropriate emotion. The limbic system is thus tightly integrated with our emotional lives. Many of its components modulate emotions and memory.

The limbic system is connected to the overlying cortex on that side of the brain. These connections are complex, and represent a major challenge in modern dementia research. The limbic system contains single regions and fibre pathways which can be studied separately. Memory is quite specific for stored material, so that, for example, verbal and visual information are handled independently. This means that within this complex circuitry one structure may be revealed to be more important in verbal memory than, say, visual or spatial memories. An example of this is the sparing of intense emotional memories in patients with severe memory problems caused by damage to a separate, discrete part of this system.

Elizabeth Warrington and Lawrence Weiskrantz in Britain took these observations to their obvious conclusion when they proposed in 1982 that the clinical problems of memory loss (including dementia) might be better understood as 'disconnection' syndromes rather than as a global phenomenon affecting several brain structures in a haphazard fashion. Here, the 'disconnection' is between parts of the integrated circuitry which support memory.

The Braaks' studies assessed the degenerative changes of Alzheimer's disease by measuring the density of neurofibrillary tangles in brain structures. In stages one and two of the disease, these are largely confined to the input area of the part of the brain known as the transentorhinal region. This structure receives its input from cortical areas on the same side of the brain. Stages three and four show more severe changes in this transentorhinal region, with extension to the next structure in the pathway, the entorhinal region. In stages five and six the changes occur throughout the cortex. The neuropathological changes in these areas may be critical determinants of memory problems. Another German neurologist, Hans Markowitsch, has suggested that the involvement of 'bottleneck' structures in the neuropathology of Alzheimer's disease is the basis of the symptoms. He argues that the anatomical basis of memory processing allows identification of critical brain regions ('bottlenecks') which are essential to memory functions, whether visual, verbal or spatial.

Signs and symptoms

Alzheimer's disease is typically of gradual onset and slow progression, in which three main phases are usually distinguished. Initially there is subjective memory impairment, poor concentration, and awareness of everyday difficulties; this phase lasts up to two years. Spatial difficulties may be typical: familiar circumstances can seem strange, route-finding is impaired and left–right agnosia may be complained of. In the second stage, symptoms suggestive of more focal cortical damage are common, although no typical pattern of deficit has been identified. The basis of these 'focal' symptoms is described in the next section. Neurological signs during the second stage may include an abnormal foot withdrawal response and some facial weaknesses. While language is preserved, it is not unusual for delusions or hallucinations to be described.

In the third stage, language is much impaired and soon lost completely, to be replaced by a continuing apathy. Many patients do not recognise themselves or their relatives. They soon become confined to bed and incontinent of both urine and faeces. Grand mal epileptic seizures are common at this stage, and neurological signs reveal gross disturbances of walking, muscle tone and features suggestive of the Klüver–Bucy syndrome (associated with bilateral temporal lobe excision). Together, the second and third stages of Alzheimer's disease's usually last about four or five years. In patients with early onset – especially those with a family history of Alzheimer's disease – the course can be much more rapid, whilst in later-onset cases the illness may last for up to twelve or even fifteen years.

The gradually progressive course of Alzheimer's disease is

associated with reduced rates of survival. Life expectancy is shortest in the early-onset form of the disease, and longer in the late-onset forms.

Confusion

The terms 'confusion' and 'delirium' are sometimes used interchangeably, and the additional term 'clouding of consciousness' makes things no clearer. Consciousness is briefly defined here as the ability to alter the focus of one's attention at will and to discriminate between environmental stimuli. It coexists with an awareness of one's own feelings and thinking. When consciousness is disturbed in confusion there is reduced distinction between competing stimuli so that the clarity of awareness typical of normal consciousness is lost. Importantly, this loss is not an 'all or none' phenomenon: there is a range of disturbed consciousness along which patients will differ, and each patient will fluctuate over time. Disturbance of consciousness is fundamental to the concept of confusion. Additional features include disorientation in time, place and person, inappropriate use of language, and an inability to complete memory tests satisfactorily. Visual phenomena are also thought to be typical of confusion and can include the misinterpretation of visual stimuli, the belief that visual experiences have occurred in the absence of an appropriate stimulus, and the sense that what is perceived is in fact something else (an illusion). Confusion usually develops quite quickly – often over a period of hours, rarely more than days. There is a marked tendency for the symptoms and signs to vary in severity and they may be interspersed with periods of complete normality.

The patient's preoccupation with their own inner world is obvious in confusion of sudden onset. In the absence of pre-existing brain disease, this inner world is often coloured by the patient's own emotional responses to the experience of confusion. Intense fear, panic and the interpretation of bizarre visual hallucinations may predominate. Confusion attributable to the ill-effects of poisoning by sedative drugs can make the patient seem apathetic or listless; however, confusion associated with withdrawal from the same sedative drugs may have pronounced features of excitation. Together, these two processes can produce an evolving clinical picture where listlessness and some features of delirium are replaced (as the toxin is excreted) by symptoms of drug withdrawal.

The effects of focal damage to the cortex

Damage to a small area of the brain's cortex can cause delirium and dementia. In Alzheimer's disease, some parts of the cortex tend to be more affected than others. Knowledge of the signs of localised ('focal') cortical damage is invaluable when assessing symptoms suggesting the onset of Alzheimer's disease. Most of this knowledge has been derived from studying people with a traumatic brain injury limited to a precise part of the cortex.

Otherwise healthy brain-damaged patients with circumscribed brain cortical lesions (usually acquired in wars or traffic accidents) have taught us a great deal about focal cortical disorders. Changes in mental ability attributable only to the lesion can be distinguished from the general consequences of brain disorder. These are conveniently divided between the frontal lobes, parietal lobes, temporal lobes and occipital lobes.

The parietal lobes

The parietal lobes are relatively small. They are often affected by Alzheimer's disease. The sensory cortex is set out in the parietal cortex with large areas for the face, arms and genitalia. Speech is located in the parietal cortex and in the upper part of the adjacent temporal lobe. The visual association area is adjacent to the occipital lobe. Lesions of either parietal lobe typically cause visuospatial and topographical difficulties. Visuospatial problems are detected in clinical examination by asking the patient to produce or copy simple drawings (such as a cube or a clock face) or to construct simple patterns from matchsticks. Difficulty in drawing something as simple as a clock face can be a very sensitive single test for early Alzheimer's disease.

Topographical problems show up when patients have difficulty in learning their way around a new environment (e.g. finding the way out), or become lost even in familiar surroundings. When the dominant parietal lobe is affected, complex defects of language appear: these include dyslexia and dysgraphia. Motor coordination is also disturbed. Non-dominant parietal lobe lesions disturb the exact awareness of body image and its relation to external space. In dementia this can include 'dressing dyspraxia' or failure to recognise familiar faces (including one's own).

Case 4

Bill D. was a fifty-three year old and worked as a driver of high-speed trains. At home he was a keen DIY enthusiast. His eldest son had just come home unexpectedly having taken leave from his college studies because of work pressures. A few days later at work Bill noticed that he seemed clumsier than usual, and dithered before the controls of his train. That evening, he was trying to finish a minor repair at home when he realised that he had forgotten which way to turn his screwdriver to fix a shelf. His family doctor could find no evidence of the little stroke he suspected, and simple testing of memory revealed no deficits. He reassured Bill and advised him to return to work. Bill felt much less confident and told his boss, who sought a second opinion from the railway doctor. Meantime, Bill was given shunting duties. The works doctor could find nothing wrong and suspected a psychiatric problem, partly because Bill's lack of confidence was so out of character and also because Bill told him he was very worried about his son. The psychiatrist heard Bill deny any difficulties but recognised the significance of his 'little problems' with the train controls and the screwdriver. More detailed psychological tests showed other symptoms to do with parietal lobe functions. A brain scan later revealed thinning of the cerebral cortex in both the parietal and temporal lobes, typical of Alzheimer's disease.

The temporal lobes

Temporal lobe tissue often degenerates in Alzheimer's disease. The numerous symptoms that ensue reflect the many functions of the temporal lobes – auditory and vestibular information processing; memory (in the hippocampus); and supplementary motor functions to do with facial expression, eating, and emotional responses to pain and pleasure.

Loss of brain cells in the dominant temporal lobe can cause language problems such as difficulty in understanding speech, loss of the ability to read or write, and certain types of difficulty in the construction of common objects. Damage to the non-dominant temporal lobe causes fewer signs or symptoms. When both temporal lobes are affected there is a devastating loss of memory, whereas unilateral lesions produce only isolated losses of certain aspects of memory function. Temporal lobe damage also causes persistent disturbances in temperament and the control of aggressive impulses. Unlike other focal brain damage, damage to the temporal lobe (especially if it extends deep into its structure) can produce a characteristic visual field defect. This is caused by damage to the fibres that convey visual information. Although it is often difficult to examine visual field defects in dementia, the presence of a defect may indicate a focal brain lesion, suggesting that the dementing illness is caused by this lesion rather than by a diffuse disease process.

Case 5

Sally E. was a stalwart at the charity shop where she was a volunteer worker. She had impeccable manners, a great sense of style and very good taste. On her seventy-second birthday her husband had surprised her with a trip to Paris which he confidently expected her to enjoy immensely. Although the journey went well she was miserable throughout the short break. Once fluent in French, she stumbled over simple phrases with their old French friends, who excused her by saying she must be tired by the journey. She seemed convinced she had offended them deeply. The next few weeks were difficult for her: she was uncharacteristically rude to other volunteers in the shop, and irritable with her husband. Then she had what later she referred to as her 'disaster'. She went to the bank to move funds from one account to another and, given forms by the counter clerk for signature, stared at them for many minutes before, flustered, leaving hurriedly without signing anything. At home she seemed overwhelmed by the events of the day. She neglected her housework and was weeping when her husband returned from golf. The family doctor wondered about depression but recognised she was unlikely to accept referral to a psychiatrist. The neurologist detected many different types of language difficulty and some memory problems. A brain scan revealed no abnormality, but the neurological diagnosis of Alzheimer's disease was confirmed as the illness slowly progressed.

The occipital lobes

All of the symptoms of occipital lobe lesions are linked to its visual functions. Typical symptoms include the inability to read but with retention of the ability to write ('alexia without agraphia'), and failure to identify colours or objects. The occipital lobes are rarely affected by Alzheimer's disease, so when these symptoms are present this diagnosis is unlikely.

The frontal lobes

Lesions of the frontal lobes produce probably the most distinctive signs and symptoms of focal cerebral disease. There is a hierarchical model of nervous system organisation which suggests that the highest mental functions (characteristics of human cognition) reside in the most recently developed parts of the cortex – the frontal lobes. Removal of frontal lobe function, in this model, produces characteristic disinhibition with over-expansive behaviour, social intrusiveness, and loss of social and moral control. Errors of judgement in all fields of human endeavour are reported. Sometimes there is a change in temperament, seen by someone who knows the sufferer well as an empty or fatuous cheerfulness without meaning or significance. Also disturbed is the ability to maintain attention and carry out sequential planned activities. Disturbance in passing urine is typical of frontal lobe lesions; it may be the first symptom and is attributed to the inability of the impaired frontal lobe to resist the urge to micturate when the bladder is full. The frontal lobes are affected by ageing to a greater extent than any other part of the brain cortex. They are sometimes involved in Alzheimer's disease, but rarely uniquely so. They are, however, singled out in certain unusual forms of dementia (frontal type dementia and Pick's disease).

Clinical examination of a person with dementia should be sufficient to establish which parts of the brain are most affected. If the clinical features point to localised brain change and there are other features suggestive of a focal lesion, then intense efforts must be made to identify a potentially treatable cause. An example would be an old person presenting with dementia mostly affecting temporal lobe functions. If additional features included a particular type of visual field defect the cause could be a slowly growing temporal lobe tumour. Removing it might be successful and reverse the dementia.

Brain scientists are now beginning to understand why brain cells shrink and die in dementia. It is not simply 'hardening of the arteries of the brain', which can explain only a small proportion of late-onset cases. In the next chapter we shall see how the new non-clinical scientists have taken up the task of disentangling the molecular pathology of Alzheimer's disease. Although many of the clues on this medical mystery trail were spotted by descriptive brain pathologists, epidemiologists and thoughtful clinicians, the story as it has unfolded from the 1980s onwards has been told by laboratory scientists.

Chapter 9

The dementia detectives

Research on the causes of dementia

The modern era of dementia research began with the publication in 1955 by Martin Roth of a comparison of outcomes for old people admitted to a large UK mental hospital immediately before and after the Second World War. Before his work was widely known, considerable therapeutic pessimism prevailed in the care of mentally infirm old people, but Roth showed that their diagnostic evaluation was a worthwhile undertaking. Effective treatment of depressive illnesses, introduced during the war, much improved the outlook for an illness which previously had a very poor prognosis in old age. Work established the importance of senile dementia as a major cause of chronic disability and death. Later, in collaboration with Bernard Tomlinson and Gary Blessed in Newcastle, Roth established a clinical research programme which successfully correlated clinical findings during dementing illnesses with brain observations after death.

Roth's work in Newcastle was greatly assisted by the development in Sweden of techniques to measure neurotransmitter biochemistry in the human brain. These techniques relied at first on the precise measurement of enzyme activity in tiny pellets of brain tissue. These enzymes were known to build up or break down specific neurotransmitters. It was then possible to take tissue from patients who had died of age-related brain degenerative diseases (such as Parkinson's disease, Alzheimer's disease and Huntington's disease) and establish which neurotransmitters were present in abnormally high or low concentrations. Once established, these neurotransmitter deficits became first the target of therapy and then a model for animal experiments. Animals were used to test if experimental toxins could replicate exactly the pattern of neurotransmitter loss found in specific brain cells in Alzheimer's disease. Later, the genetic abnormalities found in familial Alzheimer's disease (FAD) were deliberately introduced into the genetic makeup of mice to see if this could cause the typical picture of the disease.

The first treatment breakthrough based on the discovery of a neurotransmitter deficit came in the 1970s, when (contrary to most expectations at the time) neurologists working on Parkinson's disease discovered that deficits in the brain neurotransmitter dopamine could be made good by 'loading' patients with large oral doses of the dopamine precursor levodopa. This compound relieves the symptoms of Parkinsonism by replenishing the brain's stocks of dopa, especially in the substantia nigra where brain cell loss causes the typical motor features of Parkinson's disease. The human body has many natural ways of breaking down dopa before it reaches the brain; this problem is minimised by giving the levodopa in combination with an enzyme inhibitor which prevents its breakdown outside the brain but does not cross the blood–brain barrier. This combined approach allows much smaller doses to be used.

This revolution in the treatment of Parkinson's disease led dementia experts to ask if the neurotransmitter deficit in Alzheimer's disease – reported almost simultaneously by researchers in Edinburgh, Newcastle and London – could be treated in the same way. Acetylcholine was already known to be important in memory from studies on drugs that blocked its actions in the brain. The first treatment attempts relied on the administration of the drug physostigmine, which blocked the breakdown of acetylcholine. Research teams led by Ken Davis in San Francisco and Iain Glen in Edinburgh soon showed that some slight improvements in memory could be consistently produced by physostigmine in people with Alzheimer's disease. Unfortunately, the drug needed to be given intravenously along with another drug to counter its effects on the heart, and its benefits lasted for just twenty minutes or so. Nevertheless, many drug company research teams became convinced that this was the best way to go in the treatment of Alzheimer's disease. The hunt was on to find drugs that would be acceptable to patients and more effective than physostigmine.

The first compounds tried out in clinical trials looked unpromising. Tetrahydroaminacridine (tacrine) inhibited the breakdown of acetylcholine. It was effective in improving memory in about 30 per cent of patients who could tolerate the drug, and was first registered for use in the USA. Although this drug is no longer used, it taught drug researchers and government registration agencies valuable lessons about the treatment of dementia.

The next drugs in development were required to meet precise stipulations to establish their value to demented people. In addition to memory improvement measured by standard memory tests, new antidementia drugs had to meet criteria to show that they could improve the patient's quality of life and independence. This meant that – probably for the first time in drug registration – pharmaceutical companies needed to show not just reduction of the core symptoms of a target disease, but additional benefits in the life of the disease sufferer.

The next drugs followed the same treatment strategy. Galanthamine (Reminyl), rivastigmine (Exelon) and donepezil (Aricept) are currently available in Europe to treat Alzheimer's disease. They are quite similar – they all counteract the breakdown of acetylcholine – with fewer side effects than tacrine. About 60 per cent of patients treated with these drugs respond by slowing the progression of their disorder, and can maintain this response for at least nine months and perhaps for up to four years. About 10 per cent of patients will show definite improvement. This can be so dramatic that skills once thought irretrievably lost can be restored. The duration of such improvements is not yet established.

In an important sense these treatments of Alzheimer's disease are essentially symptomatic. There is some evidence that rivastigmine and donepezil may have a direct effect on the disease process, but the bulk of the evidence is that they act by preventing the breakdown of acetylcholine in the brain. The hunt is now on for drugs that can stop brain cells dying in Alzheimer's disease.

The nature of Alzheimer's disease

The association between cognitive impairment and advanced age is so well established that for many such impairment is inseparable from ageing. This view continues to have many adherents and probably underpins much of the therapeutic pessimism encountered in the management of Alzheimer's disease. For some, this disease lies at the extreme of a normal cerebral ageing continuum. This issue is not simply of academic interest. If Alzheimer's disease proves to be an abnormal variant of normal ageing then its proper study will rest firmly on the account of the neurobiology and neuropsychology of the ageing process. Alternatively, if Alzheimer's disease is a disease that is distinct from ageing, then lessons learnt from studies of well-diagnosed disease cohorts will inform the wider issue of senile dementia and provide data relevant to its treatment. The latter approach has supported the many studies of rare familial forms of early-onset Alzheimer's disease which have in turn been generalised to late-onset cases.

This question is also the subject of many community-based surveys of mental impairment in old age. In Australia, Scott Henderson and his colleagues combined the results of twenty-two surveys of dementia reported between 1945 and 1985 (this is called 'meta-analysis'). They concluded that the incidence of Alzheimer's disease approximately doubled in each five-year period after the age of seventy. In contrast, Ritchie and Kildea combined nine similar surveys and concluded that the prevalence of dementia increases exponentially from the age of sixty to age eighty, and then tends to flatten out. They commented:

> If senile dementia is considered to be ageing related – that is lying on a continuum with normal cognitive ageing – the clinical response is essentially one of palliative care for a series of progressive disabilities inevitably linked to the ageing process. On the other hand, if senile dementia is age related – that is, it is a pathological process with a specific age range being at highest risk – researchers are justified in searching for aetiological factors other than those implicated in normal ageing, with a view to providing therapeutic interventions.

Unfortunately, none of the epidemiological studies on the incidence of dementia helps to disentangle Alzheimer's disease from vascular dementia. For various technical reasons, the diagnostic criteria available introduced systematic biases which are so large as to make 'meta-analysis' a meaningless undertaking. Proper study of this important question must await the introduction and validation of new diagnostic criteria for Alzheimer's disease and vascular dementia. The question then changes. Is this disorder dependent on an ageing brain and the processes of ageing? Or is it more simply associated with ageing, so that it is age-related and not age-dependent?

So much has happened in recent research into Alzheimer's disease that it is sometimes easy to forget how the present level of interest and excitement started. What happened is one of the great medical detective stories. It has required huge collaborative scientific efforts and like all the best stories it is full of clues, some mischief, and best of all, a great deal of optimism about the future.

Chromosome 21

My first visit to the United States was at the invitation of the Kroc Foundation in 1981. Transatlantic travel was by then commonplace and most UK senior scientists had obtained their BA (Been to America) badge by the end of the 1970s. No one I spoke to in Scotland had a clue about the Kroc Foundation and they were equally mystified by the identity of its founder, the philanthropist and billionaire Ray Kroc. My brother, who had settled in Los Angeles in 1967, was better informed. 'It's McDonald's, the hamburger people,' he wrote back, while assuring me of a warm welcome to California. Despite its Scottish name, McDonald's was unknown in Scotland at that time.

The Kroc Foundation was wound up on the death of its founder, who for many years had spotted and then tried to support novel scientific research in common, chronic disabling diseases which he thought needed a little help. Ray's formula was simple and often repeated. He would issue an invitation to sixteen active research workers – always a mixture: some old, some young, some distinguished, most chosen because they had the potential to learn practical and intellectual lessons from each other. The setting was idyllic. Built on the edge of an escarpment overlooking California's Santa Ynez valley, the ranch – known locally as Old McDonald's Farm – had central conference and residential facilities for sixteen people, pools, tennis courts, a lecture room and library, and a small cinema. My six days were among the most challenging of my academic life up to that point. Acknowledged leaders attended from the fields of clinical genetics of Alzheimer's disease, the creation of gene libraries, experimental neurology and neuropathology, the molecular biology of Down's syndrome, and protein pathology.

The original idea for the meeting was suggested by Henry Wisniewski, the Polish émigré neuropathologist who had devised elegant techniques to study the abnormal proteins which accumulate in the dying and dead brain cells of Alzheimer patients. He had suggested my name to the organisers, he said, because of my relative youth (at that time) in an ageing field. I hoped he was joking; the more likely reason was the single study I had conducted on chromosomal changes in Alzheimer's dementia and ageing without dementia. My reason for starting that study – and the reason the Kroc Foundation had found so compelling – was the link between Alzheimer's disease and Down's syndrome. It took many years to show that Alzheimer's changes almost always arose in people with Down's syndrome, but no one at that time knew why. This is not really surprising because the causes of Down's syndrome were unknown for the first half of the twentieth century.

In the 1950s, Jerome Lejeune developed techniques to examine human chromosomes in health and disease. He and his colleagues in Paris spotted an oddity in Down's syndrome, namely that the normal arrangement of chromosomes was altered. Normally, there are forty-six chromosomes, arranged in pairs numbered from 1 to 23 according to size, and two sex chromosomes (X and Y). In Down's syndrome, Lejeune and his colleagues found that one of the smallest pairs (21) had an extra copy – from which he coined the term 'trisomy'. How could genes on this extra chromosome cause Alzheimer's disease? This was the question on the lips of everyone at the Kroc meeting in 1981. My task was to report a fairly straightforward study from our Edinburgh group. We had examined chromosomal arrangements in thirty-six patients with early-onset Alzheimer's disease, thirty-six non-demented subjects matched by age and sex, and thirty-six sex-matched non-demented old people some twenty years older than the Alzheimer patients. The findings were timely for the meeting. There was no evidence to suggest that chromosome 21 was involved in Alzheimer's disease. But there was evidence to suggest that chromosomal ageing was more advanced in these patients than in age-matched subjects.

At the Kroc Foundation we talked long and late about the role of chromosome 21 and related questions in Alzheimer's disease. Our discussions were made more germane by careful recent reviews of the structure and function of body systems in Down's syndrome by George Martin in Seattle. He had shown that apart from the typical facial appearance, and other features of Down's syndrome, fully apparent by late childhood, these individuals showed many characteristics of abnormal or premature ageing. Maybe, thought the scientists at the Kroc meeting, Down's syndrome is telling us something very important about the genetic control of both Alzheimer's disease and normal ageing. The working title for our meeting was 'Alzheimer's disease, Down's Syndrome and Ageing'.

Chromosome 21 is one of the smallest chromosomes. By 1981 fewer than ten individual genes had been confidently allocated to it. Pierre Sinet from Paris talked about which of these genes might cause Down's syndrome. Somehow, an extra copy of one or more of these genes (a 50 per cent increase) was enough to cause the syndrome. He speculated that a small number of these genes caused Down's syndrome, but that just one of them caused Alzheimer's disease. His evidence was intriguing. In France, Lejeune had established a laboratory to study childhood diseases and to link them with specific chromosomes. His team used techniques to make the whole chromosome seem banded from end to end – rather like the barcode on an item of shopping. Because these 'barcodes' for each chromosome were as unique as those that distinguish, say, toothpaste from toilet paper at a supermarket checkout, an experienced eye could photograph each chromosome and then line up each chromosome in pairs like criminal mugshots.

Pierre Sinet had studied some children with Down's syndrome who did not appear to have the required extra copy of chromosome 21. He had carefully examined their chromosomes and to everyone's surprise (but not his) he found a segment from the tail of chromosome 21 attached to another chromosome. Clearly, the genetic information present in this segment was enough to cause Down's syndrome; but what were the genes present here? What was needed was a detailed map of chromosome 21, especially for the region Pierre had linked to Down's syndrome. The map available at that time provided few clues.

Most interest focused on a gene coding for the enzyme superoxide dismutase 1 (SOD-1) (see Figure 2). Leonard Hayflick had already pointed to this enzyme as a possible factor in ageing. He had shown that ageing of cells could be affected by the cell's respiratory machinery producing 'free radicals' (described in Chapter 1) and implicated SOD-1 activity here. Others at the Kroc meeting gently prised these ideas apart and asked simple questions about the links between Down's syndrome, ageing and Alzheimer's disease. Everyone agreed there was a gene to be found somewhere in that pathological segment, among perhaps as few as a hundred genes. Most could see how a chromosome 21 gene library would be the place to start to look, and that perhaps some of the families multiply affected by Alzheimer's disease could help.

In the months that followed, correspondence was exchanged, molecular geneticists worked on chromosome 21 libraries, and clinical geneticists scoured the world for interested physicians who could obtain DNA from families who had at least two members with Alzheimer's disease confirmed at post-mortem. This laborious approach was rather like panning for gold, where the prospector needs little scientific knowledge but endless persistence to find the dreamt-of nugget. At the time, it seemed to be the best way forward, but better science prevailed.

Most Alzheimer's disease research teams in that period were multidisciplinary. Clinical services were not too far from the laboratory where basic scientists worked on brain material provided from post-mortem studies of patients whose relatives had consented to help combat this awful disease. Some researchers focused on the abnormal proteins that accumulate in the brain. These are of two main types. The first is attached to agglomeration of brain cell debris, the remnants of the cellular structures of dead neurons. This protein is of a type that is known to accumulate in ageing organs and because it takes up stain material much like starch, it is called 'amyloid' ('starch-like'). The amyloid in Alzheimer's disease makes up strands which align to form plaited sheets ('beta-pleated protein'). Fibrils of this protein accumulate in the brains of some old people, all patients with Alzheimer's disease, and all Down's syndrome patients after the age of forty.

A Californian scientist, George Glenner, won the race to extract and analyse the amyloid protein, reporting its structure in 1985. Like all proteins, it consisted of a chain of amino acids linked by peptide bonds. Once the amino acid sequence of a protein is known, then by a process called 'reverse genetics' it is possible to determine the structure of the DNA sequence coding for that protein. Careful screening of the gene library extracted from a human cortex found a stretch of DNA that exactly matched the amyloid sequence. This was part of a gene coding for a protein made up of about 695 amino acids, now called 'amyloid precursor protein' or APP_{695} (the subscript number refers to the number of amino acids in the protein). The protein has one section that spans the cell membrane of the neuron, with a long section extending into the extracellular space. Another section extends across part of the cell membrane and part of the intracellular space. Later research showed that APP is present in many animals other than humans (i.e. it is highly conserved in evolution), but amyloid deposition is rare in old animals. Two more varieties of APP were then detected: APP_{751} and APP_{770}.

The next step was no surprise at all: the gene for APP is on chromosome 21. It began to look as if an extra copy of the APP gene (or overexpression of this gene) would be enough to cause Alzheimer's disease in people with a normal chromosome number. In fact, a report appeared in the prestigious journal *Science* to this effect, claiming that overexpression of APP could be detected in old people with Alzheimer's disease. However, this turned out to be mistaken. The next idea was that in Alzheimer's disease there are faults in the way that the segments of APP are joined together. This proposal led to the discovery that in some families with early-onset Alzheimer's disease there are single errors in the structure of APP – in other words just one of the 695 amino acids in the chain has been replaced by another. These families are rare, and no one was sure how relevant this abnormality would be to Alzheimer's disease in old age.

This discovery of a genetic mutation in Alzheimer's disease by Alison Goate and her colleagues in John Hardy's laboratory at St Mary's Hospital, London, proved to be a pivotal point in Alzheimer's disease research. For the first time, scientists had a faulty gene to study in dementia. The obvious next step was to synthesise the faulty gene, insert it into a mouse, and then analyse how it could cause the brain changes of Alzheimer's disease. This is easily said, but takes longer to do. There were some mistakes (and some fraud) along the way. Eventually, however, researchers around the world had access to a mouse model of early-onset familial Alzheimer's disease (FAD). Abnormal APP processing could be studied, potential treatments could be examined and key questions addressed. For example, the timing of pathological changes in Alzheimer's disease was unclear. Did the amyloid plaques cause brain cell death, or were they an irrelevant consequence of dying brain cells? Did the neurofibrillary tangles disrupt normal cell function and trigger a cascade of events leading to brain cell death, or were they just the ghosts of dead neurons?

These questions were asked some ninety years after Alois Alzheimer first described a patient with characteristic lesions (plaques and tangles) in the limbic system and association cortex. At the turn of the century it seemed as though a precise map was at hand of the routes taken by brain cells from health to death. Such a map of aberrant pathways would surely, everyone hoped, provide rational treatments for dementia. Soon the causes of Alzheimer's disease would be known, and the key questions could be addressed. Is Alzheimer's disease one or many diseases? Is there a single route which always leads to Alzheimer's disease? Is there a precise pattern of brain cell loss? Are the same brain cells, the same patterns of disconnections and the same neurotransmitters always lost in Alzheimer's disease, and always in the same order?

Amyloid precursor protein

The healthy functions of APP probably include the stimulation of outgrowths from brain cells and possibly cell-to-cell recognition in the nervous system. It is now known that APP is a family of related proteins. Brain cells make APP_{695}, while outside the brain longer forms (APP_{751} and APP_{770}) are more common. These longer forms contain a segment of fifty-six amino acids which inhibits protein breakdown and is involved, for example, in the control of blood clotting. If the gene for APP is removed from the mouse, survival is unaffected and the mouse seems normal. In late life (around two years of age), mice without the APP gene are not as agile as normal mice and their brains show some slight changes, but there is no evidence of Alzheimer's disease.

In 1996, Karen Hsiao and her colleagues at the University of Minnesota created a mouse with an APP mutation copied from a Swedish family with early-onset Alzheimer's disease. The Hsiao mouse created in this way developed amyloid deposits in late life which were indistinguishable from those seen in Alzheimer's disease. These deposits did not appear in brain regions spared in Alzheimer's disease. Previous attempts to make a mouse model of Alzheimer's had failed, probably because the model did not generate sufficient amyloid for deposition to take place, or the deposits accumulated too early in life.

The Hsiao mice also showed deficits in learning and memory when compared with normal old mice, although these studies were probably too limited to provide firm conclusions. One important opportunity provided by these animal models is examination of the link between amyloid deposits and behavioural change. If the amyloid plaques must be present before the symptoms of Alzheimer's appear, then plaque formation is a likely target for treatment. However, if amyloid itself impairs brain cell function (that is, amyloid is cytotoxic) and so causes mental impairment, then 'plaque-busting' drugs would be ineffective – or even worse, these drugs could release harmful amyloid from plaques (where the brain has locked it away) to damage brain cells further. So far, it looks as though amyloid formation is alone sufficient to cause behavioural deficits in mice long before the plaques appear.

Genetic mutations in APP cause only a tiny fraction of all cases of Alzheimer's disease (certainly fewer than one in a thousand), but have taught us invaluable lessons about the role of abnormal APP processing in the disease. The APP is broken down into smaller segments by the action of enzymes (the proteases alpha, beta and gamma secretase) which cleave between amino acids (Figure 11). Beta-amyloid is generated in tiny amounts by healthy tissues and in large amounts in Alzheimer's disease. The molecule occurs in two forms, one of forty-two

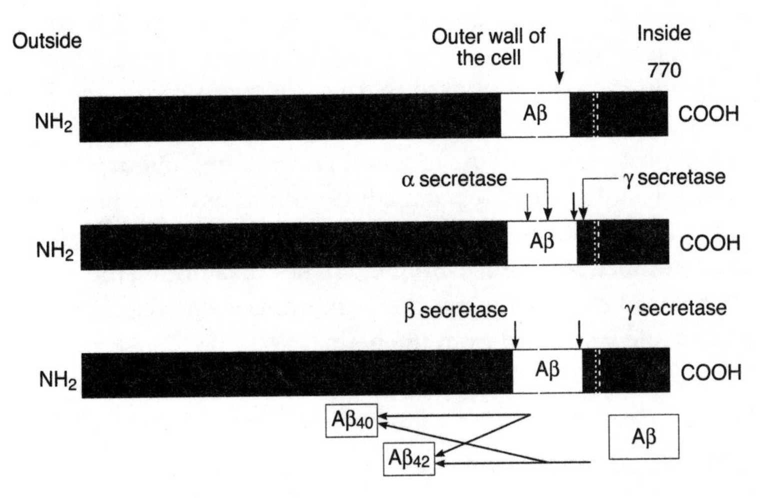

Figure 11 **The amyloid protein** is derived from a much larger protein. This larger protein is called amyloid precursor protein (APP) and is represented by the long shaded rectangle in the figure. The amyloid protein is the smaller unshaded rectangle contained within APP. Specific enzymes snip out amyloid from APP. These enzymes are called secretases. Certain rare genetic abnormalities in the regulation of these enzymes cause Alzheimer's disease. Inhibition of one type of secretase is a target for a new drug therapy; success may prevent production of the unwanted amyloid section of APP.

amino acids (Aβ42) and one of forty amino acids (Aβ40). Initially, Aβ42 is deposited in plaques largely because it aggregates into fibrils much more readily than Aβ40. Beta-amyloid accumulation is one of the earliest pathological changes in Alzheimer's disease, and could be the cause of other damage in the brain. Part of the argument for this conclusion rests on observations that even severe neurofibrillary tangle formation does not cause beta-amyloid deposition. For example, one rare type of dementia accompanied by parkinsonism destroys brain cells in the frontal and temporal cortices (frontotemporal dementia with parkinsonism, FTDP). It is linked to (and is probably caused by) a mutation of a microtubular assembly protein gene on chromosome 17 (*FTDP-17*). There are widespread neurofibrillary tangles but no amyloid deposits. Against the proposition that amyloid deposition causes Alzheimer's disease is the failure of APP mutant mice to produce neurofibrillary tangles: none of the strains studied so far has done so. Part of the puzzle is missing.

A mouse model was used in 1999 by a team from the Elan pharmaceutical company in California to test the idea that immunisation against Aβ42 could prevent amyloid deposition. Their results were encouraging: immunisation prevented amyloid deposits in young APP mutant mice and much reduced their number in old mice. Studies are now examining the effects on mouse behaviour when these immunisation techniques are used to clear amyloid from the brain.

Chromosomes 1 and 14

Once the APP mutations were recognised, families with early-onset Alzheimer's disease were screened for these mutations. It was at first disappointing that some well-known FAD pedigrees failed to show any APP mutations, but this turned out to be something of a bonus because considerable efforts were then made to determine what *was* happening in these families. These efforts uncovered unexpected links to APP processing and are now the basis for innovative treatments.

A proportion of early-onset Alzheimer's disease is caused by mutations of two genes on chromosomes 1 and 14 respectively. These genes code for proteins which are 67 per cent identical and are highly conserved in evolution. Replacements of single amino acids in either of these two proteins – presenilin 1 (chromosome 14) and presenilin 2 (chromosome 1) – are linked to familial Alzheimer's disease. Replacements occur at twenty-four sites on presenilin 1 and at two sites on presenilin 2. The proteins are made up of a series of loops, which probably determine their normal functions but which were entirely unknown when the proteins were discovered. However, gene libraries were soon systematically scrutinised for anything similar in nature and, rather surprisingly, something similar was found in the lowly roundworm, *Caenorhabditis elegans*. Here, the related protein was known to be involved in making life-or-death decisions about cells early in development. The possibility was raised that mutations of presenilin genes could be responsible for brain cell death by acting on a similar pathway in humans. Another possibility was raised by the observation that the presenilin proteins could interact with neuron-specific proteins involved in cell-to-cell recognition systems.

Neither of these suggestions proved to be the case. Mutations of the presenilin genes actually cause early-onset Alzheimer's disease by disrupting the breakdown of APP. The evidence for this is that APP mutations are located very close to the recognition sites on APP for the alpha, beta and gamma secretases. When amino acids are substituted at these sites, Aβ42 production increases and amyloid accumulates in the brain. Generation of beta-amyloid is decreased by γ-secretase which cleaves the middle of the beta-amyloid segment of APP; γ-secretase liberates beta-amyloid from the segment of APP cleaved by γ-secretase. Presenilin proteins soon became therapeutic targets in Alzheimer's disease. The search is now on for drugs that will inhibit secretases. If, as seems likely, the presenilins are essential only in early development, then their inhibition will not prove too disruptive to normal cell functions in late life. Inhibition, potentially, could prevent amyloid accumulation.

Chromosome 19

Led by Alan Roses, a team of gifted clinical neuroscientists at Duke University found a couple of fresh leads on the genetics of Alzheimer's disease. The first was that the late-onset, non-familial form of the disease, was linked to an anonymous region on chromosome 19. The second was that a transport protein – apolipoprotein E – was often found in and around the neuritic plaque, like moss on a gravestone. Apolipoprotein E (ApoE) was already known to be important in medicine – as a risk factor for vascular disease. This family of lipoproteins has an important part to play in the transport of lipids around the body. Apolipoprotein E is important in the brain and, as with many such proteins, the body will tolerate slight rearrangements of structure as long as function is not too badly compromised. The protein exists in three forms: ApoE ϵ2, ApoE ϵ3 and ApoE ϵ4. There is a corresponding gene for each form – *APOE2*, *APOE3* and *APOE4*. Alan Roses' group showed that late-onset Alzheimer's disease is linked with the ApoE ϵ4 variant. Through some unknown mechanism, possession of the *APOE4* gene increases the risk of Alzheimer's disease in old people by up to 40 per cent. It was seen as the most important single genetic factor in the disease, accounting for much more of the variation in disease susceptibility than all known APP mutations taken together.

The role of ApoE in Alzheimer's disease remains a mystery. At first it looked as if the different forms of the protein differed in the efficiency with which they helped the brain repair itself. A damaged brain cell need not die if it can be stopped from leaking its precious contents. Lipids are needed to repair brain cell membranes damaged by the actions of free radicals, and are transported into the brain by ApoE to replace lipids lost by peroxidation.

Sometimes brain cell death is inevitable. The surviving brain cells must then try to compensate for the lost cell in order to maintain function. The brain's information processing capacity does not simply depend on the number of brain cells but on the efficiency of their connections (synapses). The brain's compensatory response to brain cell loss is the formation of new synapses, a process known as 'compensatory synaptogenesis'. New synapses require new cell membranes, new sections of cytoskeleton and new presynaptic vesicles. Variants of ApoE may differ in their ability to help in synaptogenesis. Some of these processes may not directly involve lipid transport but may concern the role of ApoE in new protein synthesis. Current research is dissecting the exact role of ApoE in brain function, maintenance, repair and compensatory synaptogenesis. The promise offered by these new insights into what has remained the most important single genetic risk factor in Alzheimer's disease is that new neuroprotective drugs will be developed based on ApoE.

The association between apolipoprotein E and Alzheimer's disease was first reported in 1993. Subsequent studies have confirmed that ApoE genotype is second only to age as a risk factor for the most frequently encountered subtypes of Alzheimer's disease. About one-fifth of ApoE is produced in the brain, where it is important in the response to injury. The original study by Alan Roses' group found that the frequency of the *APOE4* allele in late-onset Alzheimer's disease patients was 50 per cent, compared with only 16 per cent in control subjects. This finding has been consistently reproduced in studies of populations of European ancestry where the *APOE4* allele occurs at a frequency of about 15 per cent.

Typically the *APOE4* allele is present in about 45 per cent of Alzheimer's disease patients. There is also a claimed association between ApoE genotype and age at onset of Alzheimer's disease: earlier-onset cases are associated with two copies of *APOE4*, whilst later-onset cases are associated with only one copy. Although the significance of this association remains unknown, there is no shortage of plausible biological hypotheses. The deposition of beta-amyloid in neuritic plaques is associated with ApoE to which it binds with high avidity. This has suggested that ApoE (ϵ4 may preferentially promote the formation of beta-amyloid fibrils by acting as a 'molecular chaperone'. Jules Poirier in Toronto has suggested an alternative hypothesis. Nerve cell repair requires internalisation of cholesterol through the ApoE/low-density lipoprotein receptor pathways; differences between ApoE variants in their affinity for cholesterol could be enough to affect the compensatory growth response of neurons. There is evidence that compensatory synaptogenesis is compromised in Alzheimer's disease.

The importance of these new data on ApoE suggests that knowledge of the ApoE genotype could be used to identify people at risk of Alzheimer's disease. In the USA, the National Institute on Ageing/Alzheimer's Association (NIA/AA) working group in 1996 considered this point and concluded that using ApoE genotype in this way would be inappropriate. Moreover, there is dispute over the use of ApoE genotype to establish a diagnosis of Alzheimer's disease. The NIA/AA group concluded that the presence of an *APOE4* allele increased the likelihood of an Alzheimer's disease diagnosis in a demented patient by as much as 14 per cent for each copy (we have up to two each). Clearly, this is no substitute for careful clinical examination and interpretation in diagnosing this disease. It may be possible in future to include the results of ApoE genetic testing in the diagnostic process, where it may prove relevant to evaluation of risk in other non-affected family members and the interpretation of responses to a putative treatment. The NIA/AA group emphasised that ApoE genetic testing should not be undertaken in the absence of appropriate resources to provide education and support, not just in the diagnostic process but in the subsequent care of the patient and the family.

The 'free radical' hypothesis of Alzheimer's disease

The association between Alzheimer's disease and ageing has suggested that the biological explanation of ageing may prove relevant to elucidating the causes of Alzheimer's disease. The 'free radical' theory of ageing would merely be of academic interest to Alzheimer's researchers were it not for the repeated observation that individuals who take non-steroidal anti-inflammatory drugs (NSAIDs) have lower rates of dementia and mental impairment in late life. The following section explains what is happening when brain cells die in Alzheimer's disease and why NSAIDs might help prevent dementia.

Beta-amyloid can be released from APP and bind to a large cell-surface molecule termed a 'receptor for advanced glycation end-products' (RAGE). Interaction between beta-amyloid and the walls of cerebral blood vessels probably causes oxidant stress. This disturbs cellular function, damages intracellular proteins and can cause cell death. In addition to their neurotoxic effects, beta-amyloid fibrils stimulate microglia to produce neurotoxins (including reactive oxygen species). In turn, these trigger a cascade of toxic events central to the neurodegenerative process in Alzheimer's disease. The interaction between Aβ fibrils and microglia is likely to be through the scavenger receptor, which probably stimulates microglia to accumulate around amyloid deposits, accounting for their proliferation in and around amyloid plaques.

The likely involvement of free radical formation in this process is interesting but not compelling. Oxidative damage to neurons in Alzheimer's disease is as likely to be a consequence of neuronal damage as a cause of it. However, oxidative stress triggered by the deposition of Aβ fibrils could contribute significantly to worsening of neuronal injury. The location within the pathological segments of chromosome 21 of the gene coding for the enzyme superoxide dismutase prompted a free radical hypothesis to account for Alzheimer's disease in Down's syndrome. Cultured neurons obtained from Down's syndrome fetal cells generate increased amounts of reactive oxygen species, presumably leading to neuronal death (apoptosis). This may account for abnormal brain development in Down's syndrome as well as the early onset of Alzheimer's disease.

The role of beta-amyloid in causing brain cell death through stimulation of microglial activity with the production of reactive oxygen species can be linked to the presence in neuritic plaques of chemical messengers called cytokines. These are released by cells when they are damaged (for example after a stroke) and orchestrate the inflammatory response. The amyloid toxicity hypothesis of Alzheimer's disease and the free radical theory of ageing are not mutually incompatible and can separately account for much Alzheimer-type degeneration in neurons. There may also be some interaction between the two processes, suggested by reports that beta-amyloid can induce free radical formation in blood vessels and also by activating neuronal membrane oxidation. Further study of the neurotoxic activity of beta-amyloid may point to therapeutic strategies (such as vitamin E supplementation of the diet) aimed at preserving cerebrovascular endothelium and reducing the neurotoxic effects.

Tau protein, neurofibrillary tangles and Alzheimer's disease

As we saw in Chapter 2, the neuropathology of Alzheimer's disease includes the formation of neurofibrillary tangles in the limbic system and linked cortical sites. These become progressively more affected as the disease progresses. The nature and distribution of neurofibrillary tangles show a close correlation with the degree of dementia. The molecular pathology of Alzheimer's disease is, therefore, more complex than would be suggested by successful navigation of the 'amyloid toxicity' hypothesis. Neurofibrillary tangles contain paired helical filaments composed of tau protein present in an abnormally phosphorylated state (Chapter 2). Tau is one of the many microtubule-associated proteins ubiquitous in nervous tissue. At least six different types of microtubule-associated protein are present in adult brain, each of which is derived from a single gene product. All six types are expressed in Alzheimer's disease. In health, tau proteins promote polymerisation of microtubules. When tau is hyperphosphorylated such polymerisation does not occur, and the protein goes on subsequently to form paired helical filaments which contribute to the characteristic lesions of Alzheimer's disease.

Neurofibrillary tangles are difficult to study, largely because of their insolubility. This feature also allows the neurofibrillary tangle to survive after the death of a neuron as a 'gravestone' or 'ghost tangle'. Early research on the molecular pathology of Alzheimer's disease focused on the senile plaques and it is only in the past few years that substantial progress has been made in understanding the molecular composition of neurofibrillary tangles. Their progressive deposition in cortical brain areas is known to follow a consistent pattern. This close relationship between pathological change and the stage of clinical dementia, together with evidence that the presence of neurofibrillary tangles provides a better distinction between healthy ageing and Alzheimer's disease than does the presence of neuritic plaques, has suggested to some researchers that the hyperphosphorylation of tau protein is the key step in the sequence of molecular events producing the characteristic neuropathological changes of Alzheimer's disease. Unravelling these molecular mechanisms may provide a therapeutic target for the development of new drugs in the prevention of Alzheimer's disease.

The glucocorticoid cascade hypothesis

The role of cortisol in brain ageing was examined earlier when it was suggested that hippocampal damage due to cortisol could lead to memory impairment and possibly Alzheimer's disease. This explanation agrees with the view that Alzheimer's disease is an extreme variant of brain ageing. Age-related changes in the brain could upset the glucocorticoid balance, leading to an exaggerated glucocorticoid response to stressors, neurotoxicity, and a vicious circle of hippocampal neuronal death. The 'glucocorticoid cascade' hypothesis was put forward by Robert Sapolsky from Harvard University. His hypothesis is particularly relevant to Alzheimer's disease because the hippocampus is so important in memory. Molecular events following the formation of beta-amyloid fibrils may perturb the function of transmembrane proteins and allow calcium entry into the cell.

Potentially, three hypotheses (for each of which there is

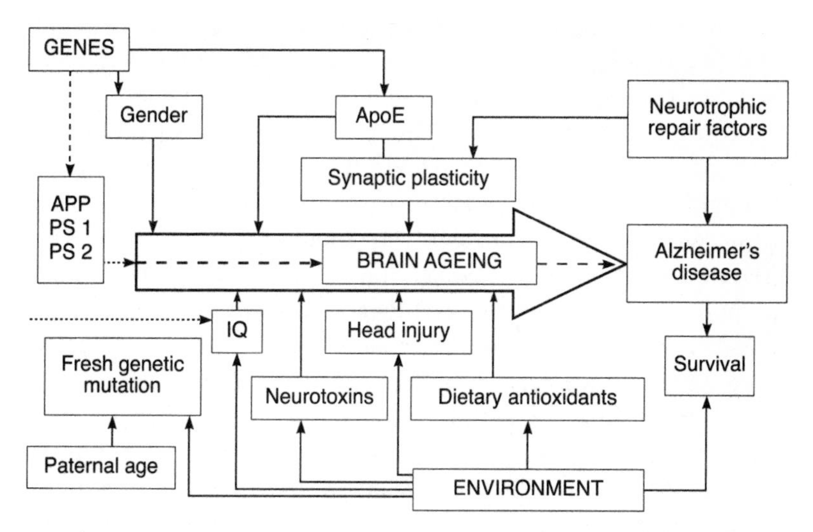

Figure 12 **Alzheimer's disease is a complex late-onset disorder.** This complex figure summarises what is important among the likely causes of Alzheimer's disease. Against a background of brain ageing (large arrow) multiple influences modify the chances of developing Alzheimer's disease. Genes are important causes but known genes account for less than 0.3 per cent of late-onset cases. ApoE does not determine *whether* or not you get Alzheimer's but *when* it starts. Previous history of head injury accounts for about 4 per cent of late-onset cases. The chances of Alzheimer's disease are greater when childhood intelligence is lower than when it is higher. The environment – especially diet and not smoking – has much potential to delay the onset of dementia.

support in studies of ageing and Alzheimer's disease) may be importantly interlinked in the causes of Alzheimer's disease. Effects attributed to the amyloid toxicity hypothesis and exacerbated according to the free radical hypothesis of Alzheimer's disease may be further exaggerated by the actions of excess glucocorticoids, as postulated in the glucocorticoid cascade hypothesis of ageing and Alzheimer's disease.

Diagnosing dementia

In middle to late age, the presence of cognitive impairment of gradual onset which follows a progressive course – in the absence of any other explanation for the cognitive impairment – suggests the diagnosis of Alzheimer's disease. The diagnosis is confirmed by the detection of characteristic lesions in the limbic system and linked cortical areas: these lesions must include brain cell loss, beta-amyloid deposits outside brain blood vessels, and the deposition of neurofibrillary tangles. These criteria (codified in 1984) are widely used in Alzheimer's disease research. Clear evidence of cerebrovascular disease (usually a stroke) before onset of cognitive impairment precludes the diagnosis of Alzheimer's disease; this excludes many individuals with 'mixed pathologies' where both cerebrovascular and Alzheimer-type changes are present, as well as people with Alzheimer's disease who have extracerebral disorders that may (or may not) have impaired cognitive function. Because more men than women suffer from cerebrovascular disease in late life there is some (as yet unquantified) systematic bias towards the over-representation of women amongst Alzheimer's disease samples. The gender bias associated with the impact of age-related disorders on the use of Alzheimer's disease diagnostic criteria is not well understood, and probably accounts only in part for the finding that Alzheimer's disease is more common in women.

The basis for dementia drug treatments

The cholinergic system – the group of neurons that use acetylcholine as a neurotransmitter – is an important component of learning and memory. Anatomical, pharmacological and neurochemical studies in ageing and Alzheimer's disease all support the idea that the functions of intact cholinergic neurons are an essential component of learning and memory. Between 1972 and 1980, such studies laid the basis for almost two decades of therapeutic research on the cholinergic system in Alzheimer's disease. At first, it appeared that reduced cholinergic function was specific to Alzheimer's disease and not found in other types of dementia. However, cross-diagnostic studies with adequate control samples found activities of enzymes associated with the synthesis (choline acetyltransferase) and degradation (acetylcholinesterase) of acetylcholine were reduced not only in Alzheimer's disease but also (although to a lesser extent) in ageing in the absence of dementia, alcoholic dementia, dementia associated with Parkinson's disease, and in some instances of vascular dementia. The conclusion that the cholinergic deficit was specific to Alzheimer's disease was soon considered unsafe.

The characteristic neurochemical changes of Alzheimer's disease may be reversed, at least in animal studies, by neurotrophic factors. Single factors acting alone are insufficient to reverse all the known neurochemical deficits, and combinations of factors may be required. Problems of neurotoxicity in drug delivery have so far barred any progress in the therapeutic use of these agents in Alzheimer's disease.

Chapter 10 Empires of the future

Better brain ageing

> *'The empires of the future will be empires of the mind'*
> *Winston Churchill (1943)*

Knut Hamsun, sometimes called the literary father of Ernest Hemingway, wrote a remarkable novel about an ordinary man in extraordinary circumstances. The novel – called *Hunger* – describes an aspiring writer's gradual starvation and subsequent descent into madness. The search for food comes to dominate the young man's life to the exclusion of all else. Starvation distorts his perceptions and then his reason. Nothing is as important as his need to find something to eat. The young man's optimism crumbles into fear and self-disgust. Hunger strips away his veneer of civilisation and he spirals into a vortex of insanity. Fatal flaws in his character are uncovered by his increasing need for food.

We can draw parallels with another sort of starvation found in extreme old age – the starvation of the senses. We are denied those experiences we cherished as adults, experiences that once so enriched our lives. Will old age bring derangement as the threat of imminent death approaches and senses fail?

Many books have been written to ease the fears of old age. Most stress the value of religious beliefs and practices; some focus on the personal foundations of wisdom and encourage self-reliance, moderation in all things and the benefits of controlled disengagement from the more hazardous aspects of life. This chapter has the less lofty ambition of considering the practical interventions that may benefit us in old age. Before these ideas can be aired satisfactorily, however, it is necessary to address a frequently asked question. What are the social consequences of large numbers of individuals living into late old age without acquiring disabilities – physical or mental – before death? This question is central to prevention of many age-related and avoidable causes of unsuccessful brain ageing. If there is a consensus that large numbers of healthy, active very old people are unwelcome, perhaps even a burden, then implementation of brain health promotion measures will be half-hearted at best and may even be obstructed in some quarters.

Detection of onset of brain ageing or dementia is an important gap in our current understanding. Without direct access to intimate brain processes this task is fraught with difficulty. For the foreseeable future, most methods of detection will rely on measures of brain performance – usually psychological tests. If attitudes change towards compulsory retirement on grounds of age, then it seems likely that checks on age-related changes in mental ability will become as routine a procedure as checks on fitness to drive. Recent research findings encourage optimism that brain ageing can be slowed and the risks of dementia substantially reduced. Most of the ideas for prevention will probably prove impractical, but the fact that so many able and energetic researchers are tackling these fundamental problems in biology generates its own optimism. There is also the prospect that computers may be able to make good age-related deficiencies in brain function and facilitate psychological development in extreme old age.

Active, ancient and nothing to do

Scientific progress may extend the life span, but will it ensure a reasonable quality of life? The question is a pressing one, especially when extensions of life span of up to forty years are promoted as possible. And how are large numbers of active, healthy old people to spend their time?

In Western developed countries, estimates of a non-disabled man aged sixty-five surviving to age eighty and remaining non-disabled in the year before death are placed at around one in four. For a non-disabled woman of the same age, the chances of surviving to age eighty-five without disability in the year before death are around one in six. This type of successful ageing is found in those who maintain an active lifestyle, eat a healthy, balanced diet, drink little alcohol, do not smoke, and continue social involvement. Although physical activity seems to be the single most important factor, each is a likely component of an overall healthy lifestyle, often well established by middle age. Geriatricians agree that if more middle-aged people adopted healthy lifestyles, many more might enjoy a healthy, successful old age free of disability.

The possibility of increased numbers of healthy 'successful agers' is not met with universal enthusiasm. Some identify economically unproductive old people as a burden on society. These views can find support among old people themselves, some of whom may long to return to useful work and may regret their retirement. Social responses to ageing tend – especially in urban societies – towards exclusion rather than involvement. Having enforced compulsory retirement on grounds of age, some communities build special environments for their 'senior citizens' which are designed to be secure and sociable, but their exclusive use by old people reinforces contemporary barriers between the old and other 'productive' sectors of society.

Less fortunate old people supported solely by social security may find access to services – even to fresh food – hampered by poor public transport and poor urban design. These social factors influence much planning for the future needs of old people and emphasise the issues underlying resistance to the prospect of increased life expectancy for everyone. What will old people do with extra years if society cannot cope with the few they already have?

One option might be to create economic roles for 'successful agers' who wish to remain at work. In Europe, only the voluntary sector has set out to recruit large numbers of successful agers to its workforce. Contrary to popular belief, it is healthy old people who do most voluntary acts. Once recruited, voluntary organisations retain the services of old people, provide proper supervision and training and obtain excellent low-cost results. Younger volunteers are mostly transient, picking up voluntary experiences as part of higher education. The same is often true of married women who use the voluntary sector as a 'stepping-stone' to full-time employment.

If voluntary organisations can tap this large potential workforce, could other paid-work sectors of the economy be more responsive to the changing age structure of society? Certainly, when skills are in short supply, some professionals can easily remain in work (full or part-time) for many years after the usual retirement age. This, however, is the exception rather than the rule. Most old people do not have the skills in short supply in an information-based economy. The likelihood must be that for the foreseeable future, old people will be restricted to work in the voluntary sector. However, advances in computer technology may provide new work opportunities for old people, freeing employers from the concern that judgements made by very old people are prone to errors linked to brain ageing. More likely is the growing tendency of employers to disregard simple calendar age and just focus on the person's capacity to do a particular job. Lessons learnt from the emergency and transport services may be generalisable to other occupations. Fitness to work may include not just simple measures of physical reserve capacity but mental reserves ('strength') as well. Novel solutions seem likely for highly skilled professionals in demanding jobs who do not wish to retire completely but would like alternative work of a less taxing nature.

Development of methods of measuring various aspects of ageing and assessing fitness to work would be welcomed by employees and employers alike. In the medical profession, hesitant steps have been taken to establish criteria of fitness for older doctors to continue to practise. So far, it seems that it is the less competent doctors who fail these tests, whilst more able doctors remain at a superior level in terms of skills and knowledge as they age. Certainly, some doctors can continue to practise competently into their eighth decade, though rarely beyond the age of about seventy-two.

Those old grey matter tests

Screening for the early detection of age-related mental decline seems more likely to be introduced for occupational rather than medical reasons. However, mental testing might be used in the future to monitor preventive strategies developed in public health medicine.

Tackling the huge public health problem of memory impairment in old age will not be possible without reliable methods of detecting mental deterioration at a very early stage. We still do not know when dementing illnesses begin, or the length of the 'lag time' between dementia onset and the first symptoms of mental decline. A common view among dementia researchers is that by the time the illness comes to the attention of the clinical services it is probably too late to arrest any fundamental disease process. Effective interventions are more likely to succeed the earlier they are introduced.

Three approaches are usually taken to detect the first symptoms of mental decline. These are psychological testing of mental performance, repeated brain scanning to detect brain cell loss, and measurement of computer-averaged brain electrical activity in response to a repeated stimulus (the averaged evoked response). Some research groups now combine these techniques, so that brain metabolic processes accompanying specific types of mental work are measured.

Psychological test performance is perhaps most often used, but interpretation is beset with difficulty, for reasons set out in Chapter 4. The best tests include a range of measures that give an estimate of general mental ability in youth, current problem-solving ability ('fluid intelligence'), spatial ability and language skills. For each of these, normal values have been defined for the general population according to specific age groups. It should be simple to compare observed scores on these psychological tests with the expected values for someone of that age, sex and background – but it is not.

The first problem is that estimates of previous mental ability are fairly crude. They provide no fine detail about the range of abilities or an individual's strengths and weaknesses and how these have changed during life. It is impossible in most countries to retrieve records of individual measures obtained at school of spatial, reasoning, numerical or planning abilities. Available methods of testing old people rely on the preservation over the adult life span of verbal ability and the good overall relationship between verbal and other abilities in large samples. The difficulties could be overcome if it were possible to access school records of mental development in youth and early adulthood rather than just examination results. The early detection of mental decline in old age would be far easier if lifetime mental attainments in a variety of abilities were recorded for everyone.

A second problem is that 'normal' scores on psychological tests vary widely around a group average. That is, the 'norm' is not as informative as one would hope. Tests that look for dementia symptoms are nearly valueless for early detection. These 'old grey matter tests' rely on the presence of quite severe symptoms. Even people with other evidence of mental slowing score at or close to the maximum, and deterioration needs to be considerable to fall outside the normal range. People who were poorly educated or were never too good at pencil and paper tests tend to do badly on this type of test, showing up as likely dementia cases when outcome studies show that they do not deteriorate further.

The best approach so far to the psychological detection of the first features of mental decline is to re-examine subjects at fixed intervals using the same mental tests. Suitable tests cover the range of abilities specified above and allow subjects not only to decline but to improve as well. Tests try to be 'culture fair' for old people. They are designed to tap into the mental processes used every day by old people to maintain their independence, and are not appropriate for testing younger people.

Brain imaging has considerable potential to detect early brain cell loss. What is not too clear is the contribution of brain cell loss to the first features of brain ageing. Some believe that the first brain change to impair mental functions in old age is the retraction of dendritic outgrowths and loss of dendritic spine density. This shrinkage could be sufficient to account for the loss in information processing speed, memory and spatial ability decrements observed in the first stages of dementia. Current brain imaging techniques do not detect loss of spine density, but new techniques to measure brain work during the performance of specific tasks hold some promise. These techniques rely on functional magnetic resonance imaging (fMRI) in real time, which has the potential to provide very high spatial resolution of the brain at work.

The averaged evoked response has been used in many cohort studies to chart the progress of dementia. Several research groups are refining this technique to detect the earliest sign of decline. By combining this method with fMRI some researchers expect to have a simple, safe test for early dementia very soon.

Can we repair the brain?

In both the two main types of dementia, Alzheimer's disease and vascular dementia, brain cells die and mental functions are subsequently impaired. Lessons from stroke are often applied to vascular dementia. These show that when the blood supply to part of the brain is blocked, brain cells in the worst affected areas die immediately from lack of oxygen: this is 'ischaemic' cell death. The worst affected area is surrounded by a region where brain cells do not die immediately but may do so a little later – the 'penumbra' of the stroke's ischaemic area. The likely explanation is that the chemical signals released by dying cells at the stroke's core trigger the cell suicide programmes in cells of the penumbra. Cell death of this type is called apoptosis and is controlled by enzymes (caspases) which cleave large protein molecules. Caspase inhibitors appear to have potential for the treatment of stroke and (by extension) vascular dementia, if it is accepted that these share common disease mechanisms. At the other end of the age spectrum, this class of drug may even be helpful in prevention of brain damage in infants caused by oxygen starvation during a difficult birth. David Hotzman of Washington University, Missouri, has already tested this idea in newborn rats and believes that caspase inhibitors may help prevent tragedies such as cerebral palsy.

Fats in the brain are broken down by free radicals in a process called membrane lipid peroxidation. Many age-related neurodegenerative conditions including Alzheimer's disease and vascular dementia show evidence of this type of very messy breakdown. Fats are essential components of every cell membrane in the body, and brain cells are not spared in this process. Antioxidant compounds (such as vitamin E) will prevent neuronal death by stopping membrane lipid peroxidation.

Apoptosis is a much neater form of cell death than that following membrane lipid peroxidation. Brain cells first shrink in an orderly way, then die. The cell first unpacks its DNA from the cell nucleus and then wraps up portions of DNA in envelopes of membrane. These do not trigger an inflammatory response. Unfortunately, cell death in the stroke penumbra has some but not all features of apoptosis: the DNA looks as if it has been packaged differently from a typical apoptotic cell. Nancy

Rothwell at Manchester University runs a large research programme to examine cell death in stroke and is interested in its relevance to vascular dementia, ageing and Alzheimer's disease. She is carefully taking apart the chemical signals released during brain cell death, how these signals can be blocked, and the benefits of doing so in terms of brain cell survival after stroke. So far, results look promising.

It is a well-known phenomenon in brain development that some cells are genetically programmed to die after ensuring or guiding the development of other brain cells destined to make up the adult brain. Programmed cell death of this type involves apoptosis. The observation that beta-amyloid can induce apoptosis in brain cell cultures makes apoptosis a popular theory of cell death in Alzheimer's disease. This idea is strengthened by detection of some of the key molecular components of apoptosis in brain cells examined after death in Alzheimer's disease; these include the characteristic DNA breaks and packaging.

A role for apoptosis in Alzheimer's disease was first suggested by Gianluigi Forloni in Milan and Carl Cotman at the University of California at Irvine near Los Angeles. They showed that beta-amyloid triggered a genetic programme inside brain cells which led to the cell dismantling itself as described above. Rudolf Tanzi's research group at Harvard University has explored this phenomenon further; they feel confident that natural presenilins are anti-apoptotic whereas the mutant forms found in Alzheimer's disease actually accelerate the process. In healthy individuals, during apoptosis, the caspases degrade the presenilins so that these naturally occurring inhibitors of apoptosis are neutralised. Mutant forms of presenilin, however, are more easily degraded by the caspases. Potentially, a drug strategy designed to prevent presenilins being broken down by caspases could slow down apoptosis after stroke and in Alzheimer's disease. How far down the road of cell death these events might occur is very uncertain. The likelihood is that they occur at the point of origin of the 'mixed' form of dementia. In this condition, which accounts for about 15 per cent of all dementias in old age, Alzheimer's changes coexist with those of vascular dementia.

As long ago as the 1970s, Anthony Cerami in New York proposed that sugars could influence body ageing processes. In

Chapter 2, the products of sugars attached to large regulatory biomolecules (advanced glycation end-product, abbreviated to AGE) were identified as one of the mechanisms involved in tissue ageing. Brain ageing might be slowed if AGE formation could be retarded or reversed. The obvious route would be to reduce sugar intake and this may be the reason why restricting the diet delays ageing. By 1990, Cerami had discovered compounds that slow AGE formation. These discoveries derived from something that the processed food industry had known about since 1920: the addition of sulphites to foods keeps them fresh by inhibiting sugar–protein bonding. By deliberately targeting the sugar–protein bond, Cerami discovered that a chemical called aminoguanidine helped diabetic rats maintain healthy blood vessel and kidney function. More recent work by Cerami and his colleagues suggests that compounds may soon be discovered that will break down AGEs after they have formed. One such possible compound is demethyl-3-phenacylthiazonium chloride. In animals, this compound will reverse the ravages caused by AGE formation in diabetes. The promise is a new generation of drugs designed to counter the harmful chemical changes of ageing in many tissues, including the brain.

So far we have talked about brain cell death in general. This does not explain the selective pattern of loss seen in ageing, and to a much greater extent in Alzheimer's disease. Could this selective pattern be saying something important about the origins of mental decline?

Cholinergic neurons which project from deeper structures onto the cortex are selectively lost in Alzheimer's disease. Might it be possible to 'rescue' these cells by the use of neurotrophins? These are highly potent molecules made in the brain and involved in the differentiation, maintenance and repair of neurons. When mature brain cells are damaged they can be saved from certain death by the local application of a specific neurotrophin. There is now much research into the structure, function and mechanisms of action of these molecules. One idea about the selective nature of brain cell loss in ageing and Alzheimer's disease is that the brain cells most affected have lost their neurotrophic factor, either because it is no longer available or because they cannot bind to it anymore.

Cholinergic brain cells are the best-understood population of cells that respond to nerve growth factor (NGF). Treatment of Alzheimer's disease with this neurotrophin therefore seems an obvious strategy, but attempts were beset with difficulty. These difficulties are worth a brief rehearsal, because they temper some ill-thought-out enthusiasms. The most serious concerns drug delivery. Nerve growth factor is a large molecule which binds to specific receptors on the surface of cholinergic cells. Simple infusion into the bloodstream will not deliver the NGF to the part of the brain where it is required: the blood–brain barrier blocks its way, and there are plenty of digestive enzymes *en route* to chew it up. An alternative is to inject the NGF into the cerebrospinal fluid, which bathes the brain and spinal cord and can be accessed at the base of the spine or (heroically) at the back of the skull. When administered in this way, NGF circulates over the surface of the brain and then percolates down through its substance. The highest concentrations are near the surface of the brain. The target is the basal forebrain projection of cholinergic neurons most damaged in Alzheimer's disease; instead, the NGF is given to the entire brain in quite unpredictable amounts, probably disrupting the fine balance that must exist between the various types of neurotrophin. This is entirely different from the situation in nature, where neurotrophic factors are released from adjacent support cells in the nervous system. The receptors for NGF are so widespread throughout the brain that a great deal of it must be 'mopped up' before it reaches the cholinergic target.

Although this approach has so far been unsuccessful, improvements in methods of drug delivery to selected brain areas are sure to be made. However, a more promising avenue is the development of drugs to mimic the actions of specific neurotrophins. These drugs might be attached to a neutral substance by a chemical bond that is specifically broken by an enzyme present only at the intended site of drug action. Several enzymes involved in brain cell acetylcholine manufacture and breakdown are being investigated with this in mind.

Will gene therapy help?

Selective neuronal loss occurs in other neurodegenerative diseases, including Parkinson's disease (loss of dopamine cells of the substantia nigra) and motor neuron disease. Neurotrophic factors have been seen as potential therapies for both diseases, but most early work has been done on motor neuron disease. The first discovery was that insulin-like growth factor 1 promotes motor neuron survival, as do ciliary neurotrophic factor and brain neurotrophic growth factor. All three proved either unhelpful or less effective than expected in motor neuron disease. Continuous administration of neurotrophic factors by gene therapy may prove to be more successful. Here, the patient receives an intramuscular injection of the gene for the neurotrophic factor attached to a virus that will target selected populations of brain cells. The overall success of this approach to neurodegenerative disease depends ultimately on knowing the precise requirements of specific brain cells in terms of the cocktail of neurotrophic factors and micronutrients needed to ensure their survival.

In Alzheimer's disease, gene therapy has progressed slowly. The 'amyloid toxicity' hypothesis has been discussed as a potential target. One idea is to inhibit the expression of the amyloid precursor protein (APP) gene by using small molecules (peptides) to block the processing of APP (to release Aβ42) after it has been synthesised from its gene. Another is to block expression in brain of the APP gene altogether. Blocking gene expression in this way (either totally or partially) is achievable using 'antisense oligonucleotides' – small segments of synthetic genetic material. As with the neurotrophic factors, there are huge problems delivering the drug to the required site in the brain. Many doubt the therapeutic potential of this approach.

Can brain cell death in dementia be reversed?

The idea of brain repair may seem like science fiction but is being taken very seriously. Prestigious scientific journals have published thoughtful reviews from distinguished neuroscientists on the chances of success. At first inspection, the task seems impossible. Just how could something as complex as the brain –

even just a small part of it – be reconstructed? The answers rely on progress in understanding how the brain is built up in early development, how it maintains the specialisation of its terminally differentiated neurons, how these are repaired in life, and how the brain heals itself after injury.

Studies of brain cell transplants in animals have shown repeatedly that developmental mechanisms once thought unique to fetal and early life are present in the adult mammalian brain. The pioneering experiments were reported in 1985 by Mary Kay Floeter and Ted Jones. The problem faced by workers following up these observations is that the adult brain develops mechanisms to hold in check any further brain development. The technique sometimes called 'brain self-healing stimulation' probably holds more promise than attempts to graft new cells into a diseased brain. It avoids problems of tissue rejection and inadvertent infection with a brain virus, as well as the taxing ethical issues associated with using cells from an aborted human fetus. However, the tissue transplantation technique based on the cloning technology used to create Dolly the sheep may prove valuable; here nucleated cells from the affected person would be 'reprogrammed' and induced to differentiate into precursor neurons, which would then be implanted into the damaged area of brain.

Studies on the regeneration of brain cells after injury suggest that the brain may have a much greater capacity for self-repair than was once thought by scientists like Santiago Ramon y Cajal – although we should acknowledge the foresight he showed towards the end of his life in writing about the need for scientific progress to break down barriers to mending nervous tissue. In some mammals, brain cells continue to divide and make new connections throughout adult life; some of the brain structures involved are known to be important to memory. Fetal precursor cells transplanted into an adult rat brain migrate and make functional connections, partly helped by the synthesis of guidance molecules throughout adult life. The fact that these molecules continue to be made when the need for their instructions is thought long past poses the question: just what is going on in the healthy brain? Without recourse to transplantation other scientists have shown that the mammalian brain contains 'dormant' precursor brain cells, with the exciting possibility that these cells may retain the potential to be activated and make

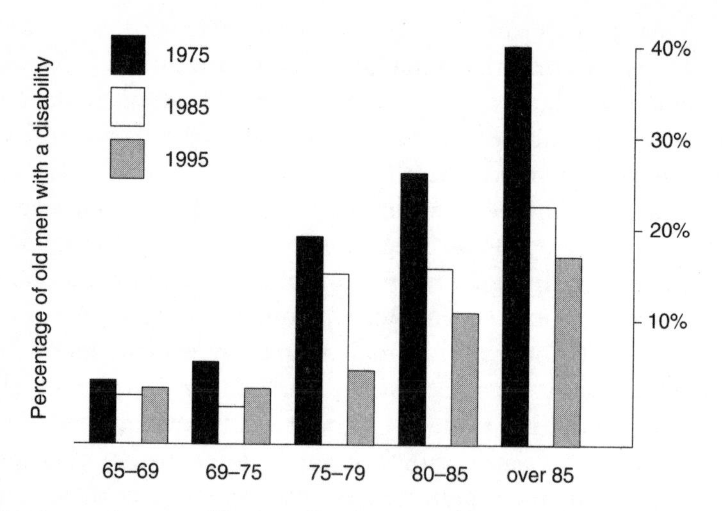

Figure 13 **Sources of brain cells of potential value** and the guidance of these cells to sites where they can make good the losses of dementia are shown to the left of this figure (a). To the right is shown how brain cells slither along processes laid down by guidance cells to reach their destination (b).

good the damage caused by injury or neurodegeneration. Other evidence points to the ability of precursor brain cells present in the mature mammalian nervous system to make functional connections as far away as the other side of the brain. As in the developing nervous system, glial cells help these outgrowths by provision of a surface along which they may slither. We need to find out how to stimulate dormant cells into becoming functional neurons, and how to stem the harmful responses of some glial cells to brain reconstruction.

The richness of brain cell connections accounts for much of the brain's complexity. It is this complexity that makes the task of reversal of brain cell death in dementia seem so daunting.

All this may seem pretty academic and of doubtful relevance to the care of people with dementia. One leading researcher on corticogenesis, Pasko Rakic, does not think so. In his commentary on these landmark discoveries, he observed:

> The findings also provoke new ideas for therapeutic strategies. For instance, repulsive molecules might be used to prevent metastasis (the spread of cancer cells) in the brain, or to streamline transplanted cells into parts of the brain affected by neurodegenerative disorders.

It takes only a little imagination, however, to speculate that brain cells might be grown outside the body to replace those lost by brain ageing or illness. The problem then becomes one of ensuring that these implanted cells take up their exact positions and make the connections needed to restore brain function. The correct synthetic molecules could be used to 'stream' transplanted cells engineered to 'read' the address message. The molecular knowledge needed by the developing human infant to build an adult nervous system would become relevant to repairing the ravages of dementia or restoring the ageing brain.

Synaptic plasticity

The molecular determinants of synaptic plasticity (the ability of neurons to alter their structure and function) are the very building blocks of perception, memory and reasoning. Neuroscientists have discovered hundreds of different molecular mechanisms involved in synaptic plasticity, such as the switching on or off of individual genes, the modification of enzymes that make or degrade neurotransmitters, the actions of neurotrophic factors, and many others. But this is far from being the complete picture. We are only beginning to understand the molecular basis of synaptic plasticity.

Individual synaptic plastic mechanisms have never been linked to any specific mental process such as an act of reasoning or reflection. However, the developmental biology of the nervous system suggests a close relationship between the reduction of excess neurons, synapses and axons in childhood and adolescence and the appearance of higher mental abilities. As the brain cell stops dividing and acquires its final shape and position, so it experiences a gradual but substantial restriction in choice.

It is lucky for brain scientists that the molecules involved in the regulation of brain function are similar to (in some cases the same as) molecules that regulate cells outside the nervous system. Common to many body systems are molecules that regulate gene expression, cell-to-cell signalling, cell differentiation, and cell growth and death. This is not to say that the brain is no different from other organs such as the kidney, but rather to emphasise that the complexity of brain processes relies on the ability of brain cells to weave their highly evolved, large regulatory biomolecules into novel and complex patterns.

These facts drive a great deal of current research in brain science. It does not seem unreasonable to assume that similar brain processes are involved in memory and reasoning, and share similar molecular mechanisms. However, knowledge is limited. It is certainly unsafe to assume that, once discovered, a particular molecular mechanism might explain all forms of a certain type of memory or its disorders. All lessons learnt so far point to considerable diversity. Some mechanisms of synaptic plasticity rely on changes in the expression of genes for neurotrophic factors; others alter the expression of genes for neurotrophic factor receptors. In other words, there is no single molecular mechanism of synaptic plasticity. Indeed, carefully separated modules of brain function probably rely on entirely distinct molecular mechanisms. In the face of such complexity, it may seem overoptimistic to try to define detailed aspects of the underlying molecular mechanisms.

The biology of synaptic plasticity is closely linked in several theoretical models of brain function to the molecular basis of memory and general mental ability. Some evidence points to these biological effects persisting throughout the life span. In 1997, Gerald McClearn at Pennsylvania State University and Bob Plomin at the Institute of Psychiatry in London provided some evidence for a large genetic contribution to mental abilities in a study of Swedish twins aged over eighty. The apolipoprotein E gene is but one gene known to influence mental decline in late life. In the ageing brain the same molecular genetic mechanisms may be recruited to mount the 'compensatory' response to loss of dendritic spines and eventual brain cell loss. If so, individual differences in the biological basis of general mental ability may contribute to differences in susceptibility to dementia. This intriguing possibility is so far untested. It is distinct from the usual model of brain cell death in brain ageing and awaits the discovery of precise genetic contributions to the mental ability of children. At this point two separate threads of research in ageing brain science may interlock. Weaving those strands into a new tapestry of understanding is a challenge ahead. Successful or not, progress in the molecular biology of the ageing brain will touch all of us who are lucky enough to grow old.

Castles of enchantment, palaces of wisdom, or fortresses of solitude

Environmental influences shape the complex patterns of connections between brain cells. These are the basis of memory and retention and this capacity is fundamental to the compensatory response to brain ageing. Could the environment be redesigned with very old people in mind? Elements known to promote successful brain ageing, for example activity and social involvement, are already incorporated into designs for assisted living spaces for old people. But what about the continuing development of mental ability, memory, judgement and so on? These present much more of a challenge, but have prompted the design of 'intelligent' assisted living spaces which 'teach' residents the way around, stimulate language use, and give aural cues to position and reminders about self-care. These designs are essentially works of control engineering in which the internal environment adjusts for the presence of the resident and advises when specific boundaries are crossed. They go nowhere near exploiting the full potential of modern computing systems. Computing science can do much more than provide a tool for environmental control: it offers a paradigm that should prove especially beneficial to old people, especially those who pioneer the fourth phase of mental development – very late active old age.

The current generation of old people do not take easily to computers. They find the keyboard unfamiliar and the monitor difficult to read. However, the middle-aged are more confident about computers, and most youngsters are comfortable working with standard software packages. The advance of computer technology, and a drive towards making computers cheaper and easier to use, strengthens the impression that new generations of portable personal computers will help people cope with ageing in the future. The drive towards computers that are easier to use and powerful enough that part of their processing capacity can be set aside just to make them 'friendly' is strongly linked to commercial imperatives. Only forty years ago computers relied on punched cards to read data or instructions, later replaced by the keyboard and then by more natural interfaces like the car controls in driving simulators. Myron Kreuger pioneered the development of computer interfaces that appear to surround the operator – the 'virtual world'. Computer graphic designers (using sophisticated image generation and processing) fed television screens mounted inside a binocular headset. A motion detector picked up head movements and instructed the computer to re-evaluate the images in keeping with the head position. By these means the phenomenon of parallax was introduced, which helped to create the sensation that the observer is *within* the 'virtual' world.

Old people do not take easily to this type of simulated experience. The computer takes too long to recalculate images with changing position so the result is jerky and crude, and in old people may induce slight symptoms of motion sickness (especially nausea). The limitations of present-day technology should not lead us to condemn the whole idea of computer-generated simulation of reality (SOR) for use by old people. Probable improvements in imaging techniques will eventually enable old people to interact simply and efficiently with computers. But for what purpose?

The 'Turing test' of computer technology demands that conversation generated by a computer cannot be distinguished by a trained observer from conversation with an unseen participant. Old people are likely to find this type of conversation with a home computer helpful and at times companionable. Simple reminders, reading text and explaining articles about recent news events could be attractive to some. More appealing to many would be the retention by the computer of an intelligently organised store of visual and auditory materials from the old person's past. 'Nostalgia machines' of this type could play a useful role in the mental imagery and conversation of old people. When the capacity to converse is linked to an accessible archive of material from a person's past and there is the option to immerse the subject into that archive, SOR machines would find an easy application in the recreational life of old people. It is then a small step to set aside some of the huge power of an advanced computer to retain day-to-day memories from the current life and through questions and answers provide reliable factual accounts which make good some slight lapse of memory. Although unfamiliar and intrusive in terms of existing technology, the personal support offered by these developments is as likely to revolutionise our lives as did the personal computer of the 1980s. Critics of the development of SOR machines for old people often focus on the artificiality of the reality created, which is only as valid as the experience of the computer graphics designer. This Disneyfication of experience might render the experiences less meaningful, more trite and be actively avoided by some very old people.

Advances in computer design and function are inevitable and will lead to computers providing more and better solutions to problems in the control engineering of food production, energy use, security and so on. These advances contain the potential for some types of computer to make decisions with far-reaching consequences for the survival of certain groups of individuals. Some science fiction writers have even envisaged computer decisions about competitive survival of subgroups of people in a way that would be preferred to human judgements on the same question; that is, computers are expected to show the wisdom of Solomon concerning perhaps the distribution of scarce resources at some distant time in the future. The potential ethical consequences are vast if we allow intelligent machines not simply to support the day-to-day lives of old people but to decide if those lives should continue in preference, say, to jeopardising the future of younger people.

The 'enchanted castle' with images and experiences created within the processing capacity of an advanced computer could become a seductive trap, holding little of any true value and relieving society at large of any responsibility to explore the new frontiers of human development opened by increased life expectancy. The parallel could be drawn with the plight of healthy adults, conscious but dependent on permanent life support and immersed for recreational purposes in 'experience machines'. How much time should the intensive care staff allow patients to spend in such machines? When is it preferable to spend all the waking day inside rather than outside in a world where the patient is totally dependent?

Advanced computers have another role in the development of mental imagery and progress in human creativity, one that seems likely to affect greatly mental development in the 'fourth age'. In addition to simulating reality, computers may also simulate complexity. Here, the computer becomes like the classroom of a gifted teacher, who uses simple materials to make the complex seem simple. The term 'envisioning' is sometimes used in computer graphics design to describe how computers can be used in this way. What is most remarkable about these developments concerns how close the logic of scientific enquiry and creativity comes to an artistic vocabulary. Scientists and artists now find common ground as the new computer graphics made possible by advances in computer design facilitate better understanding of complex problems. What happens as galaxies collide or inside a nuclear explosion are issues beyond the grasp of most of us, but computer-generated simulations may bring them within our comprehension. The reality devised by these means opens many doors of understanding. In Chapter 3, the 'theatre' metaphor was introduced as an easy model of how the brain views external reality. The outside world is brought through the mechanisms of the senses onto a stage viewed by modules of brain cells (the 'unconscious audience' framework). Computer-generated simulations of reality are examples of this metaphor; the insider is looking out onto the world. But what if the computer can generate models we can understand of more complex issues, perhaps allowing us to look deeper inside ourselves? Does this then become the outsider looking inwards? Just what will be seen?

The scene is set to use computers to develop educational programmes for the very old which explore and extend our capacity for insight, to create mental models of complex problems, rearrange elements in the spaces of our minds, and propose novel solutions. In figurative terms, it is as though we have been seated in Plato's cave, our backs to the daylight, interpreting the outside world from the patterns made by the flicker of the fire on a darkened, distant wall.

What will old people make of all this? If we assume that society is sensible enough not to develop computers that may rationally decide that old people are unnecessary, then simulators of complexity probably provide the best chance of using the extra years of life expectancy to further our mental development. We could then envisage very old people whose biological integrity in the face of great age depended upon advances in molecular biology, and whose psychological health relied on the continuing quest for knowledge and deeper understanding. Perhaps, in these extra years gifted by science, single individuals will come to understand themselves and each other more completely than ever before in our brief history.

Further Reading

Berg, J.M., Karlinsky, H., Holland, A.J. *Alzheimer's Disease, Down's Syndrome and their Relationship* (Oxford University Press, Oxford, 1993).

Bergener, M. *Psycho-Geriatrics. An International Handbook* (Spring Publishing Company, 1988).

Billig, N. *To Be Old and Sad. Understanding Depression in the Elderly* (Lexington Books, 1987).

Bourlière, F. *The Assessment of Biological Age in Man* (World Health Organization, Geneva, 1970).

Bromley, D.B. *The Psychology of Human Ageing* (Harmondsworth, Penguin, 1966).

Callahan, D. *What Kind of Life? The Limits of Medical Progress* (Simon and Schuster, New York, 1990).

Coons, D.H. *Specialized Dementia Care Units* (The Johns Hopkins University Press, 1991).

Cooper, E.L. *Stress, Immunity, and Aging*. Immunology Series, Volume 24. (Marcel Dekker, Inc., 1984).

Erikson, E.H. *Childhood and Society* (Paladin, London, 1977).

Evans, J.G., Williams, T.F. *Oxford Textbook of Geriatric Medicine* (Oxford Medical Publications, 1992).

European Symposium on Education and Older Adults: A Mirror to Society (European Education Network, Volkhogenschool Ons Erf, Berg en Dal, 1991).

Finch, C.E., Johnson, T.E. *Molecular Biology of Aging*. UCLA Symposia on Molecular and Cellular Biology, New Series, Volume 123 (Wiley-Liss, New York, 1990).

Fuchs, V. *Who Shall Live?* (Basic Books, 1974).

Gazzaniga, M.S. *The Cognitive Neurosciences*. A Bradford Book (The MIT Press, 1995).

Houle, C.O. *The Inquiring Mind*. (University of Wisconsin Press, Madison, 1961).

Hunter, S. *Dementia. Challenges and New Directions* (Jessica Kingsley Publishers, London, 1997).

Iqbal, K., Mortimer, J.A., Winblad, B., Wisniewski, H.M. *Research Advances in Alzheimer's Disease and Related Disorders* (John Wiley & Sons, 1995).

Kandel, E.R., Schwartz, J.H., Jessell, T.M. *Essentials of Neural Science and Behavior* (Prentice Hall International, Inc., International Edition, 1995).

Kaplan, H.I., Sadock, B.J. *Concise Textbook of Clinical Psychiatry.* Derived from Kaplan and Sadock's *Synopsis of Psychiatry*, 7th edn (Williams and Wilkins, 1996).

Katona, C., Levy, R. *Delusions and Hallucinations in Old Age* (Gaskell, Royal College of Psychiatrists, 1992).

Kelner, K.L., Koshland, D.E. (eds) *Molecules to Models: Advances in Neuroscience.* Papers from *Science* 1986–9 (American Association for the Advancement of Science, 1989).

Kerwin, R. *Neurobiology and Psychiatry.* Cambridge Medical Reviews, Volume 2 (Cambridge University Press, 1993).

Kirkwood, T. *Time of Our Lives* (Weidenfeld & Nicolson, London, 1999).

Kolb, B., Wishaw, I.Q. *Fundamentals of Human Neuropsychology*, 3rd edn (W.H. Freeman and Company, New York, 1990).

Levi, L. *Society, Stress and Disease.* Volume 5. Old Age. (Oxford Medical Publications, 1987).

Light, L.L., Burke, D.M. *Language, Memory, and Aging* (Cambridge University Press, Cambridge, 1988).

Lishman, W.A. *Organic Psychiatry. The Psychological Consequences of Cerebral Disorder* (Blackwell Scientific Publications, 1987).

Lovestone, S., Howard, R. *Depression in Elderly People* (Martin Dunitz, London, 1996).

Medawar, P. *An Unsolved Problem of Biology* (H.K. Lewis, London, 1952).

Minois, G. *History of Old Age: From Antiquity to the Renaissance*, translated by S.H. Tenison (Chicago University Press, 1989).

Moody, H.R. *Ethics in an Ageing Society* (Johns Hopkins University Press, Baltimore, 1992).

Patten, J. *Neurological Differential Diagnosis*, 2nd edn (Springer, 1996).

Percy, K. 'Opinions, Facts and Hypotheses: Older Adults and Participation in Learning Activities in the United Kingdom', in Glendenning, F., Percy, K. (eds), *Ageing, Education and Society*, (Association for Educational Gerontology, Keele, 1996).

Plomin, R., DeFries, J.C., McClearn, G.E., Rutter, M. *Behavioural Genetics*, 3rd edn (Freeman, 1980).

Porter, R. *The Greatest Benefit to Mankind. A Medical History of Humanity from Antiquity to the Present* (Fontana Press, 1997).

Ramon Y Cajal, S. *1852–1937. Recollections of my Life*, translated by E.H. Craigie (MIT Press, Cambridge, 1989).

Robinson, D. *The Mind* (Oxford University Press, Oxford, 1998).
Rose, M.R. *Evolutionary Biology of Ageing* (Oxford University Press, Oxford 1991).
Rosenberg, M. *Society and the Adolescent Self-Image* (Princeton University Press, Princeton, 1965).
Sapolsky, R.M. *The Ageing Brain and Mechanisms of Neurone Death* (MIT Press, Cambridge, 1992).
Schaie, K.W. *Intellectual Development in Adulthood* (Cambridge University Press, Cambridge, 1996).
Sloane, P.D., Mathew, L.J. *Dementia Units in Long-Term Care* (The Johns Hopkins University Press, 1991).
Smith, C.U.M. *Elements of Molecular Neurobiology*, 2nd edn. (John Wiley & Sons, 1996)
Smyer, M.A., Qualls, S.H. *Understanding Aging. Ageing and Mental Health* (Blackwell Publishers, 1999).
Tallis, R. *Increasing Longevity: Medical, Social and Political Implications* (Royal College of Physicians of London, London, 1998).
Timiris, P.S., Bittar, E.D. *Advances in Cell Aging and Gerontology* (JAI Press Inc, 1997).
Wang, E., Snyder, D.S. *Handbook of The Aging Brain* (Academic Press, 1998).
Wolf, Jr., S.G., Finestone, A.J. *Health and Performance at Work. Occupational Stress* (PSG Publishing Company, Inc., 1986).
Woodruff-Pak, D.S. *The Neuropsychology of Aging* (Blackwell Publishers, 1997).
World Bank *Averting the Old Age Crisis: Policies to Protect the Old and Promote Growth* (Oxford University Press, London, 1994).
Young, M., Schuller, T. *Life after Work: The Arrival of the Ageless Society* (Harper Collins, London, 1991).

Selected references

Abramson, L.Y., Seligman, M.E.P. and Teasdale, J.D. 'Learned Helplessness in Humans'. *Journal of Abnormal Psychology* 87: 49–74 (1978).

Albert, M.S., Savage, C.R., Blazer, D., et al. 'Predictors of cognitive change in older persons: MacArthur studies of successful aging'. *Psychology and Aging* 10(4): 578–89 (1995).

Arbuckle, T.Y., Maag, U., Pushkar, D., Chaikelson, J.S. 'Individual differences in trajectory of intellectual development over forty-five years of adulthood'. *Psychology and Aging* 13: 663–75, (1988).

Ashkenazi, A., Dixit, V.M. 'Death receptors: signaling and modulation'. *Science* 281: 1305–8 (1998).

Baars, B.J. 'Metaphors of consciousness and attention in the brain. *Trends in Neuroscience* 21: 58–62 (1998).

Bachman, D.L., Wolf, P.A., Linn, R.T., et al. 'Incidence of dementia and probable Alzheimer's disease in a general population: The Framingham study'. *Neurology* 43: 515–19 (1993).

Ball, M. 'Neuronal loss, neurofibrillary tangles, and granulovacuolar degeneration in the hippocampus with ageing and dementia'. *Acta Neuropathology* 37: 111–18 (1977).

Barger, S.W., Harmon, A.D. 'Microglia activiation by Alzheimer amyloid precursor protein and modulation by apolipoprotein E.' *Nature* 388: 878–81 (1997).

Barinaga, M. 'Is apoptosis key in Alzheimer's disease?' *Science* 281: 1303–4 (1998).

Barinaga, M. 'New imaging methods provide a better view into the brain'. *Science* 176: 1974–6 (1997).

Barker, D.J., Osmond, C. 'Death rates from stroke in England and Wales predicted from past maternal mortality'. *British Medical Journal* 295: 83–6 (1987).

Behan, D.P., Heinrichs, S.C., Troncoso, J.C., et al. 'Displacement of corticotrophin releasing factor from its binding protein as a possible treatment for Alzheimer's disease'. *Nature.* 378: 284–7 (1995).

Bellamy, D. 'Assessing biological age'. *Gerontology* 41: 32–324 (1995).

Beyreuther, K., Masters, C.L. 'Serpents on the road to dementia and death'. *Nature Medicine* 3: 723–5 (1997).

Blessed, G., Tomlinson, B.E., Roth, M. 'The association between quantitative measures of dementia and of senile change in the cerebral gray matter of elderly subjects'. *British Journal of Psychiatry*, 114: 797–811 (1968).

Bowling, A., Grundy, E., Farquhar, M. 'Changes in network composition among the very old living in inner London'. *Journal of Cross-Cultural Gerontology* 10: 331–47 (1995).

Braak, H., Braak E. 'Staging of AD's related neurofibrillary changes'. *Neurobiology of Aging* 16(3): 271–8 (1995).

Brookes, A.J., St Clair, D. 'Synuclein proteins and Alzheimer's disease'. *Trends in Neurological Sciences* 17: 404–5 (1994).

Bruce-Jones, P.N., Crome, P., Kalra, L. 'Indomethacin and cognitive function in healthy elderly volunteers'. *British Journal of Clinical Pharmacology* 38: 45–51 (1994).

Busciglio, J., Yankner, B.A. 'Apoptosis and increased generation of reactive oxygen species in Down's syndrome neurons in vitro'. *Nature* 378: 776–9 (1995).

Caplan, L.R. 'New therapies for stroke'. *Archives of Neurology* 54: 1222–4 (1997).

Carp, A., Peterson, R., Roelfs, P. 'Adult learning: interests and experiences', in Cross, Valley (eds.), *Planning Non-traditional Programs: An Analysis of the Issues for Postsecondary Education* (Jossey Bass, San Francisco, 1974).

Christensen, H., Korten, A.E., Jorm, A.F., et al. 'Education and decline in cognitive performance: compensatory but not protective'. *International Journal of Geriatric Psychiatry* 12: 323–30 (1997).

Clayton, D.F., George, J.M. 'The synucleins: a family of proteins involved in synaptic function, plasticity, neurodegeneration and disease'. *Trends in Neurological Sciences* 21: 249–54 (1998).

Cobb, J.L., Wolf, P.A., Au, R., et al. 'The effect of education on the incidence of dementia and Alzheimer's disease in the Framingham study'. *Neurology* 45: 1707–11 (1995).

Corder, E., Saunders, A., Risch, N., et al. 'Protective effect of apolipoprotein E type 2 allele for late onset Alzheimer's disease'. *Nature Genetics* 7: 180–3 (1994).

Corder, E., Saunders, A., Strittmatter, W., et al. 'Gene dose of apolipoprotein E type 4 allele and the risk of Alzheimer's disease in late onset families'. *Science* 261: 921–3 (1993).

David II., J.N., Chisholm, J.C. 'The "amyloid cascade hypothesis" of AD: decoy or real McCoy?' *Trends in Neurological Sciences* 20: 558–9 (1997).

Davies, S.J.A., Silver, J. 'Adult axon regeneration in adult CNS white matter', *Trends in Neurological Sciences* 21: 515 (1998).

De Strooper, B., Saftig, P., Craessaerts, K., et al. 'Deficiency of presinilin 1 inhibits the normal cleavage of amyloid precursor protein'. *Nature* 391: 387–90 (1998).

Deary, I.J., Whalley, L.J., Lemmon, H., Crawford, J.R., Starr, J.M. 'The stability of individual differences in mental ability from childhood to old age: follow up of the 1932 Scottish mental survey'. *Intelligence* 27: 1–7 (2000).

Eames, M., Ben-Shlomo, Y., Marmot, M.G. 'Social deprivation and premature mortality: regional comparison across England'. *British Medical Journal* 307: 1097–101 (1993).

Eccles, M., Clarke, J., Livingstone, M., Freemantle, N., Mason, J. 'North of England evidence based guidelines development project: guideline for primary care management of dementia'. *British Medical Journal* 317: 802–8 (1998).

Elsayed, M., Ismail, A.H., Young, R.J. 'Intellectual differences of adult men related to age and physical fitness before and after an exercise program'. *Journal of Gerontology* 35: 383 (1980).

Engert, F., Bonhoeffer, T. 'Dendritic spine changes associated with hippocampal long-term synaptic plasticity'. *Nature* 399: 66–70 (1999).

Eriksson, P.S., et al. 'Neurogenesis in the adult hippocampus'. *Nature Medicine* 4: 1313–17 (1998).

Farrer, L.A., Cupples, A., van Duijn, et al. 'ApoE genotype in patients with Alzheimer's disease: implications for risk of dementia among relatives'. *Annals of Neurology* 38: 797–808 (1995).

Fratiglioni, L., Ahlbohm, A., Viitanen, M., et al. 'Risk factors for late onset Alzheimer's disease: a population-based case-control study'. *Annals of Neurology* 33: 258–66 (1993).

Freund, A.M., Baltes, P.B. 'Selection, optimization, and compensation as strategies of life management'. *Psychology and Aging* 13: 531–43 (1998).

Freund, A.M., Smith, J. 'Content and function of the self-definition in old and very old age'. *Journals of Gerontology, Psychological Sciences and Social Sciences* 54: 55–67 (1999).

Galanis, D.J., Petrovitch, H., Launer, L.J., et al. 'Smoking history and middle age and subsequent cognitive performance in elderly Japanese-American men. The Honolulu-Asia Aging Study'. *American Journal of Epidemiology* 145: 507–15 (1997).

Gale, R., Martyn, C.N., Cooper, C. 'Cognitive impairment and mortality in a cohort of elderly people'. *British Medical Journal* 312: 608–11 (1996).

Gao, S., Hendrie, H.C., Hall, K.S., et al. 'Relationships between age, sex, and the incidence of dementia and Alzheimer disease'. *Archives of General Psychiatry* 55: 809–15 (1998).

Gillman, M.W., Cupples, L.A., Gagon, D., et al. 'Protective effect of fruit and vegetables on development of stroke in men'. *Journal of the American Medical Association* 273: 1113–17 (1995).

Goate, A., Chartier-Harlin, M.C., Mullan, M., et al. 'Segregation of a missense mutation in the amyloid precursor gene with familial Alzheimer disease'. *Nature* 349: 704–6 (1991).

Goedert, M. 'Tau protein and the neurofibrillary pathology of Alzheimer's disease'. *Trends in Neurological Sciences* 16: 460–65 (1993).

Gorman, W.F., Campbell, J.D. 'Mental acuity of the normal elderly'. *Journal of the Oklahoma State Medical Assocation* 88: 119–23 (1995).

Gould, E., et al. 'Learning enhances adult neurogenesis in the hippocampal formation'. *Nature Neuroscience* 2: 260–5 (1999).

Harman, D.A. 'Theory based on free radical and radiation chemistry'. *Journal of Gerontology* 11: 298–300 (1956).

Harrington, C.R., Colaco, C.A.L.S. 'Glycation of tau protein. Implications for the aetiopathogenesis of Alzheimer's disease'. in *Microtubule-Associated Proteins: Modifications in Disease* (Harwood Academic Publishers 1997).

Harrison, P.J. 'S182: from worm sperm to Alzheimer's disease'. *The Lancet* 346: 388 (1995).

Henderson, A.S. 'The epidemiology of Alzheimer's disease'. *British Medical Bulletin* 42: 3–10 (1986).

Henderson, A.S., Easteal, S., Jorm, A.F., et al. 'Apolipoprotein E allele (4, dementia, and cognitive decline in a population sample'. *The Lancet* 346: 1387–90 (1995).

Henderson, A.S., Easteal, S., Jorm, A.F., Mackinnon, A.J., Korten, A.E., Christensen, H., Croft, L., Jacomb, P.A. 'Apolipoprotein E allele (4, dementia, and cognitive decline in a population sample'). *The Lancet* 346: 1387–1390 (1995).

Henderson, A.S., Jorm, A.F., Korten, A.E., et al. 'Environmental risk factors for Alzheimer's disease: their relationship to age of onset and to familial or sporadic types'. *Psychological Medicine* 22: 429–36 (1992).

Jick, H., Zornberg, G.L., Jick, S.S., Seshadri, S., Drachman, D.A. 'Statins and the risk of dementia'. *The Lancet* 356: 1627–1631 (1995).

Kirkwood, B.L., Wolff, S.P. 'The biological basis of ageing'. *Age and Ageing* 24: 167–71 (1995).

Koch, C., Laurent, G. 'Complexity and the nervous system'. *Science* 284: 96–8 (1999).

Kohonen, T., Hari, R. 'Where the abstract feature maps of the brain might come from'. *Trends in Neurological Sciences* 22: 135–9 (1999).

Korten, A.E., Jorm, A.F., Jiao, Z., Letenneeur, L., Jacomb, P.A., Henderson, A.S., Christensen, H., Rogers, B. 'Health, cognitive and psychosocial factors as predictors of mortality in an elderly community sample'. *Journal of Epidemiology and Community Health* 53: 83–8 (1999).

Kuypers, J.A. 'Internal-external locus of control, ego functioning and personality characteristics in old age'. *The Gerontologist* 12, 168–73 (1971).

La Rue, A., Koehler, K.M., Wayne, S., et al. 'Nutritional status and cognitive functioning in a normally aging sample: a 6 year reassessment'. *American Journal of Clinical Nutrition* 65, 20–29 (1997).

Lamberts, S.W.J., van den Beld, A.W., van der Lely, A-J. 'The endocrinology of aging'. *Science*, 278: 419–24 (1997).

Lees, K.R. 'If I had a stroke . . .'. 'Stroke' supplement to *The Lancet*, October 1998, 352: sIII, 28–30 (1998).

Lowenstein, D.H., Parent, J.M. 'Brain, heal thyself'. *Science* 283: 1073–1216 (1999).

Magistretti, P.J., Pellerin, L., Rothman, D.L., et al. 'Energy on demand'. *Science* 283: 496–7 (1999).

Maletic-Savatic, M., Malinow, R., Svoboda, K. 'Rapid dendritic morphogenesis in CA1 hippocampal dendrites induced by synaptic activity'. *Science* 283: 1805–8 (1999).

Mann, S. 'Wearable computing: a first step towards personal imaging'. *Computer* 30: 25–32 (1997).

Manton, K.G., Corder, L., Stallard, E. 'Chronic disability trends in elderly United States populations: 1982–1994'. *Proceedings of the National Academy of Sciences USA* 94: 2593–8 (1997).

Mattson, M.P. 'Modification of ion homeostasis by lipid peroxidation: roles in neuronal degeneration and adaptive plasticity'. *Trends in Neuroscience* 21: 53–7 (1998).

Mattson, M.P., Carney, J.W., Butterfield, D.A. 'A tombstone in Alzheimer's?' *Nature* 373: 481 (1995).

McClearn, G.E., Johansson, B., Berg, S., Pedersen, N.L., Ahern, F., Petrill, S.A., Plomin, R. 'Substantial genetic influence on cognitive abilities in twins 80 or more years old'. *Science* 276: 1465–1612 (1997).

McKay, R. 'Stem cells in the central nervous system'. *Science* 276: 66–71 (1997).

Molloy, D.W., Beerschoten, D.A., Borrie, M.J., et al. 'Acute effects of exercise on neuropsychological function in elderly subjects'. *Journal of the American Geriatrics Society* 36: 348 (1989).

Morgan, K. 'The Nottingham longitudinal study of activity and ageing: a methodological overview'. *Age and Ageing* 27(Suppl. 3): 5–11 (1998).

Morris, J.Z., Tissenbaum, H.A., Ruvkun, G. 'A phosphatidylinositol-3-oH kinase family member regulating longevity and diapause in *Caenorhabditis elegans*'. *Nature* 382: 536–9 (1996).

Morrison, J.H., Hof, P.R. 'Life and death of neurons in the aging brain'. *Science* 278: 412–19 (1997).

Nalbantoglu, J., Gilfix, B.M., Bertrand, Y., et al. 'Predictive value of apolipoprotein E genotyping in Alzheimer's disease: results of an autopsy series and an analysis of several combined studies'. *Annals of Neurology* 36: 889–95 (1994).

Osuntokun, B.O., Ogunnyiy, A.O., Lekwauwa, G.U., et al. 'Epidemiology of age-related dementias in the Third World and aetological clues to Alzheimer's disease'. *Tropical and Geographical Medicine* 43: 345–51 (1991).

Ott, A., Stolk, R.P., van Harskamp, F., Pols, H.A.P., Hofman, A., Breteler, M.M.B. 'Diabetes mellitus and the risk of dementia. The Rotterdam Study'. *Neurology* 53: 1937–42 (1999).

Pentland, A.P. 'Smart rooms'. *Scientific American* 274: 68–76, (1996).

Perrig, W.J., Perrig, P., Stähelin, H.B. 'The relation between antioxidants and memory performance in the old and very old'. *Journal of the American Geriatrics Society* 45: 718–25 (1997).

Peto, R., Doll, R. 'There is no such thing as ageing: old age is associated with disease, but does not cause it'. *British Medical Journal* 315: 1030–32 (1997).

Plassman, B.L., Welsh, K.A., Helms, M. 'Intelligence and education as predictors of cognitive state in late lilfe: a 50 year follow-up'. *Neurology* 45: 1446–9 (1995).

Poirier, J. 'Apolipoprotein E in animal models of CNS injury and in Alzheimer's disease'. *Trends in Neurological Sciences* 17: 525–30 (1994).

Poirier, J., Davignon, J. Bouthillier, D., et al. 'Apolipoprotein E polymorphism and Alzheimer's disease'. *The Lancet* 342: 697–9 (1993).

Prescott, P., Primatesta, P. *Health survey for England '96* (The Stationery Office, London, 1998).

Prince, M.J., Birt, A.S., Blizzard, R.A., et al. 'Is the cognitive function of older patients affected by antihypertensive treatment? Results from 54 months of the Medical Research Council's treatment trial of hypertension in older adults'. *British Medical Journal* 312: 801–3 (1996).

Rakic, P. 'Neurobiology: discriminating migration'. *Nature* 400: 315–16 (1999).

Richardson, J.T.E., Rossan, S. 'Age limitations and the efficacy of imagery mnemonic instructions'. *Journal of Mental Imagery* 18 (3&4): 151–64 (1994).

Riddle, D.L. 'A message from the gonads'. *Nature* 399: 308–9 (1999).

Ritchie, K., Kildea, D. 'Is senile dementia "age-related" or "ageing-related"? – evidence from meta-analysis of dementia prevalence in the oldest old'. *The Lancet* 346: 931–4 (1995).

Rockwood, K., Ebly, E., Hachinsky, V., et al. 'Presence and treatment of vascular risk factors in patients with vascular cognitive impairment'. *Archives of Neurology* 54: 33–9 (1997).

Rogers, S.L., Friedhoff, L.T. 'Long-term efficacy and safety of donepezil in the treatment of Alzheimer's disease: an interim analysis of the results of a US multicentre open label extension study'. *European Neuropsychopharmacology* 8: 67–75 (1998).

Roman, G.C., Tate Michi, T.K., Erkinjuntti, T., et al. 'Vascular dementia diagnostic criteria for research studies: report on the NINDS-ARIEN International Workshop'. *Neurology* 43(2): 250–60 (1993).

Rosen, W.G., Terry, R.D., Fuld, P.A., et al. 'Pathological verification of ischaemic score in differentiation of dementias'. *Annal of Neurology* 7: 486–8 (1980).

Rosenberg, R.N. 'A causal role for amyloid in Alzheimer's disease: the end of the beginning; 1993 American Academy of Neurology Presidential Address'. *Neurology,* 43: 851–6 (1993).

Rosler, M., Anand, R., Cicin-Sain, A., et al. 'Efficacy and safety of rivastigmine in patients with Alzheimer's disease: international randomised controlled trial'. *British Medical Journal* 318: 633–40 (1999).

Roth, G.S., Joseph, J.A., Preston Mason, R. 'Membrane alterations as causes of impaired signal transduction in Alzheimer's disease and aging'. *Trends in Neurological Sciences* 18: 203–6 (1995).

Rozzini, R., Ferrucci, L., Losonczy, K., et al. 'Protective effect of chronic NSAID use on cognitive decline in older persons'. *Journal of the American Geriatrics Society* 44: 1025–29 (1996).

Sano, M., Ernesto, C., Thomas, R.G., et al. 'A controlled trial of selegiline, alpha-tocopherol, or both as treatment for Alzheimer's disease'. *New England Journal of Medicine* 336: 1216–23 (1997).

Scheffler, B., Horn, M., Blumcke, I., et al. 'Marrow-mindedness: a perspective on neuropoiesis'. *Trends in Neurological Sciences* 22: 348–57 (1999).

Schenk, D., Barbour, R., Dunn, W., Gordon, G., Grajeda, H., Guido, T., et al. 'Immunization with amyloid attenuates Alzheimer-disease-like pathology in the PDAPP mouse'. *Nature* 400: 173–7 (1999).

Schmidt, R., Fazekas, F., Reinhart, B., et al. 'Estrogen replacement therapy in older women: a neuropsychological and brain MRI study'. *Journal of the American Geriatrics Society* 44: 1307–13 (1996).

Schoenberg, B.S., Kokmen, E., Okazaki, H. 'Alzheimer's disease and other dementing illnesses in a defined United States population: incidence rates and clinical features'. *Annals of Neurology* 22(6): 724–9 (1987).

Starr, J.M., Whalley, L.J. 'Drug-induced dementia: incidence, prevention and management'. *Drug Safety* 11: 310–17.

Starr, J.M., Whalley, L.J., Deary, I.J. 'The effects of antihypertensive treatment on cognitive function: results from the HOPE study'. *Journal of the American Geriatrics Society* 44: 411–15 (1996).

Starr, J.M., Thomas, B.M., Whalley, L.J. 'Population risk factors for hospitalisation of stroke in Scotland'. *International Journal of Epidemiology* 25: 276–81 (1996).

Starr, J.M., Thomas, B.M., Whalley, L.J. 'Familial or sporadic clusters of presenile Alzheimer's disease in Scotland: I. parental causes of death in Alzheimer and vascular presenile dementias'. *Psychiatric Genetics* 7: 141–6 (1997).

Starr, J.M., Thomas, B.M., Whalley, L.J. 'Familial or sporadic clusters of presenile Alzheimer's disease in Scotland: II. case kinship'. *Psychiatric Genetics* 7: 147–52 (1997).

Starr, J.M. 'Blood pressure and cognitive decline in the elderly'. *Current Opinion in Nephrology and Hypertension* 8: 347–51 (1999).

Starr, J.M., Deary, I.J., Lemmon, H., Whalley, L.J. 'Mental ability age 11 years and health status at 77 years'. *Age and Ageing* 29: 523–8 (2000).

Thomas, B.M., McGonigal, G., McQuade, C.A., Starr, J.M., Whalley, L.J. 'Survival in presenile dementia: effects of urbanization and socioeconomic deprivation'. *Neuroepidemiology* 16: 134–40 (1997).

Thompson, E.E., Krause, N. 'Living alone and neighborhood characteristics as predictors of social support in late life'. *Journal of Gerontology: Social Sciences* 53B: S354–S364 (1998).

Tilburg, T. van. 'Losing and gaining in old age: changes in personal network size and social support in a longitudinal study'. *Journals of Gerontology: Social Sciences* 53B: S313–S323 (1998).

Tolkovsky, A. 'Neurotrophic factors in action – new dogs and new tricks'. *Trends in Neurological Sciences* 20: 1–3 (1997).

Vargha-Khadem, F., Gadian, D.G., Watkins, K.E., Connelly, A., van Paesschen, W., Mishkin, M. 'Differential effects of early hippocampal pathology on episodic and semantic memory'. *Science* 277: 376–80 (1997).

Vogel, G. 'Gene discovery offers tentative clues to Parkinson's'. *Science* 276: 1973 (1997).

Van Praag, H., et al. 'Running increases cell proliferation and neurogenesis in the adult mouse dentate gyrus'. *Nature Neuroscience* 2: 266–70 (1999).

Vaughan, C.J., Delanty, N. 'Neuroprotective properties of statins in cerebral ischaemia and stroke'. *Stroke* 30: 1969–73 (1999).

Von Dras, D.D., Blumenthal, H.T. 'Dementia of the aged: disease or atypical-accelerated ageing? Biopathological and psychological perspectives'. *Journal of the American Geriatrics Society* 40: 285–94 (1992).

Wardell, F., Lishman, J., Whalley, L.J. 'Volunteers: making a difference?' *Practice*, 9(2): 21–34 (1997).

Whalley, L.J., Gerhand, S. 'What to measure in dementia' in Nimmo, W., Tucker, G. (eds.), *Clinical Measurement in Drug Evaluation* (John Wiley & Sons, 1995), 23–43.

Whalley, L.J., Thomas, B.M., Starr, J.M., McGonigal, G., McQuade, C.A., Black, R. 'Migration and risk factors for Alzheimer's presenile dementia in Scotland' Iqbal, K., Mortimer, J., Windblad, B., Wisniewski, H.M. (eds.), *Research Advances in Alzheimer's Disease and Related Disorders* (John Wiley & Sons Ltd, 1955), 31–41.

Whalley, L.J. 'Ethical issues in the application of virtual reality to medicine'. *Computers in Biology and Medicine* 25: 107–14 (1995).

Whalley, L.J., Thomas, B.M., McGonigal, G., McQuade, C.A., Swingler, R., Black, R. 'Epidemiology of presenile Alzheimer's disease in Scotland (1974–88): (1) non-random geographical variation'. *British Journal of Psychiatry* 167: 728–31 (1995).

Whalley, L.J., Thomas, B.M., Starr, J.M. 'Epidemiology of presenile Alzheimer's disease in Scotland (1974–88): (2) exposures to possible risk factors'. *British Journal of Psychiatry* 167: 732–8 (1995).

Whalley, L.J., Struth, M. 'The prediction of cognitive decline in late life (review)'. *Alzheimer Research* 3: 177–89 (1997).

Whalley, L.J., Struth, M. 'The prevention of cognitive decline in late life (review)'. *Alzheimer Research* 3: 261–73 (1997).

Whalley, L.J. 'Can dementia be prevented?' *Practitioner* 242: 34–8 (1998).

Whalley, L.J. 'Vascular dementia (review)'. *Proceedings of the Royal College of Physicians of Edinburgh* 28: 602–10 (1998).

Whalley, L.J., Starr, J.M., Athawes, R., Hunter, D., Pattie, A., Deary, I.J. 'Childhood mental ability and dementia'. *Neurology* 55: 1455–9 (2000).

Whalley, L.J. 'Early onset Alzheimer's disease in Scotland: a review of genetic epidemiological studies'. *British Journal of Psychiatry* (2001).

Whalley, L.J., Deary, I.J. 'Longitudinal Cohort Study of Childhood IQ and Survival up to age 76'. British Medical Journal 322: 819–22 (2001).

Winkler, J., Suhr, S.T., Gage, F.H., et al. 'Essential role of neocortical acetylcholine in spatial memory'. *Nature* 375: 484–7 (1995).

Winocur, G., Moscovitch, M., Freedman, J. 'An investigation of cognitive function in relation to psychosocial variables in institutionalized old people'. *Canadian Journal of Psychology* 41(2): 257–69 (1987).

Wisniewski, A., Frangione, B. 'Apolipoprotein E: a pathological chaperone protein in patients with cerebral and systemic amyloid' *Neuroscience Letters* 135: 235–8 (1992).

Wu, W., Wong, K., Chen, J., Jiang, Z., Dupuis, S., Wu, J.Y., Rao, Y. 'Directional guidance of neuronal migration in the olfactory system by the protein Slit'. *Nature* 400: 331–6 (1999).

Index